YOUR LIFE, SOLVED

42 Practical Ways AI Can Help You Save Time, Reduce Stress, and Get More Done

MARLON BUCHANAN

HomeTechHacker.com

ISBN: 978-1-958648-07-0 (paperback)
ISBN: 978-1-958648-08-7 (hard cover)
ISBN: 978-1-958648-10-0 (ebook)
ISBN: 978-1-958648-09-4 (audiobook)

Edited by Peter Letzelter

CONTENTS

Other Books by Marlon Buchanan

- *Life by Design: Automate to Master Time, Health, Money, and Unlock Personal Success*
- *The Personal Cybersecurity Manual: How Anyone Can Protect Themselves from Fraud, Identity Theft, and Other Cybercrimes*
- *The Home Network Manual: The Complete Guide to Setting Up, Upgrading, and Securing Your Home Network*
- *The Smart Home Manual: How to Automate Your Home to Keep Your Family Entertained, Comfortable, and Safe*

All of these books are available in eBook, paperback, and audiobook formats from all major retailers and directly from my website. You can learn more about them at MarlonBuchanan.com.

Subscribe to HomeTechHacker.com

Be sure to check out HomeTechHacker.com and subscribe to my newsletter for free in-depth articles about using artificial intelligence, home automation, home networks, cybersecurity, and more. In addition to the articles, you'll find the HomeTechHacker Technology Advisor, an easy-to-use tool that provides personalized advice for home technology purchases in under a minute.

PREFACE

Technological change has been exponential in our lifetimes. From the mass adoption of the personal computer to the rise of the web to smartphones, it seems that our day-to-day lives have been changing every decade. Artificial intelligence (AI) is the latest technology revolutionizing our lives.

One of the problems with rapid technological change is that it is hard to keep up with. For many, it feels like you have to be a tech expert just to keep up with the tools, terms, and capabilities. AI has increased that tenfold. I consider myself a technology expert; at times, I even find all the incoming information about AI to be overwhelming.

However, AI doesn't have to be overwhelming. The key is to focus on what the technology can do for you. The point of technological advances is to make our lives easier and better, to enable us to do things we couldn't previously. I've learned that the way to keep from becoming overwhelmed is to focus on the ways AI can make my daily life better.

The goal of this book is to show you practical and simple ways to leverage AI to solve many of the day-to-day tasks and frustrations of life. You'll learn how to use AI tools and techniques to tame the decision fatigue we all face, how to lighten the mental load of life, how to get everyday tasks done faster, and how to level up your home, relationships, and work performance.

Through practical tips, real-world examples, and my own experiences, I'll guide you step by step on a journey that covers how to use AI to improve your life. It's about working smarter—not harder—and making technology your ally in every area of life.

Many of the techniques described in this book were also part of how the book itself was created. During the writing process, I used AI tools as thinking partners, research assistants, and editors to help organize ideas, explore examples, and refine explanations. Every concept, recommendation, and final decision in this book reflects my own judgment and experience, but AI played a role in helping shape and improve the material. In many ways, writing this book became its own real-world experiment in using AI to solve problems and improve the way we work.

I hope this book provides inspiration to embrace AI, as well as the ability to apply it to your life in practical ways. Let's turn AI from hype into a partner.

INTRODUCTION

Artificial intelligence is inescapable. It is all over the news, social media, workplaces, and discussions with friends and family. Technology in general can feel overwhelming, but AI seems to be in its own category.

For many people, especially busy professionals and parents, technology and the other burdens of everyday modern life are already overwhelming. AI may feel like yet another technology that you'll have a hard time keeping up with. AI, like many tech tools, can be more of a time suck than something helpful.

Though it may seem like it, AI isn't just hype or a fad. It's a key tool that can enrich your life in ways you may not realize. However, you have to work through the overwhelming AI buzz and cut to the tools and techniques that will benefit you—now and for years to come.

That's what this book is about.

Not the AI buzz, or the details of how AI works, or even an exhaustive listing of all the AI tools out there that you could be using. There will be no doom-and-gloom stories about how AI will eventually destroy the world. Instead, this book focuses on real-world practical applications, written in plain words that anyone can apply to their own life.

At the risk of sounding cliché, you only have one life to live. One that is filled with challenges, responsibilities, and complications. I'm here to tell you that AI can help with many of life's obstacles.

When you finish this book, you'll know how to use AI to do the following:

- Eliminate decision fatigue, manage time better, and get rid of the feeling of being overwhelmed.
- Put everyday annoyances and tasks on autopilot so you can focus on things that will benefit most from your attention.
- Improve work productivity and boost creativity.
- Develop better relationships with family, friends, and coworkers.
- Help your family with their goals and responsibilities.
- Level up your health and fitness.

This book is your go-to guide for using artificial intelligence to make life easier, more organized, and way less stressful. Its 42 real-world challenges provide the know-how to put AI to work for you.

HOW TO USE THIS BOOK

Each of the 42 ways that AI can be used to solve common life problems follows a consistent structure. First, I define the problem clearly, then I explain why it persists. Next, I show how specific AI tools can help. I close with real-world results and short takeaways to remember. At the end of each chapter, I tie the ideas together and provide a list of the tools referenced. The title of each chapter is an example of a problem the chapter will solve with everyday AI tools.

This book makes AI practical and personal by blending self-help with real-world AI technology applications. Although it is intended to be read sequentially, it works fine as a reference for those who want to jump to particular aspects of how artificial intelligence can solve a life problem. Be aware, though, that some sections build on information presented earlier. If jumping around, you may need to reference other sections to take full advantage of the solution you are targeting. Fortunately, the book's layout and the descriptive headings make it easy to find information.

Be sure to take advantage of the additional resources in the appendix and glossary. These resources supplement the skills gained from the book on the journey to enhance your life with AI skills.

TOOLS YOU'LL MEET

This book has discussions and real-world examples of accessible AI tools. You'll see uses of ChatGPT, Copilot, and Claude. Other popular tools I cover include Notion AI, Notebook LLM, Perplexity, and Gemini. You'll also be exposed to some lesser-known but easy-to-use AI tools.

Throughout the book, I reference ChatGPT because it's widely known and easy to access. You can use whichever large language model (LLM) you prefer—Claude, Gemini, Grok, Copilot, or others—and expect similar results. The specific tool matters less than how the input is structured and the output utilized. The book's principles and workflows are model-agnostic. Use the platform that best fits your ecosystem and privacy preferences.

CONVENTIONS USED IN THIS BOOK

Here are conventions I use, which highlight important information:

- **My personal experience**: In several sections, I give specific information by addressing topics in the context of my own life. In the section about building a custom AI assistant, for example, I discuss how I've used AI assistants for fitness, life coaching, and business advising.

- **Takeaways**: Each section ends with a single-sentence, plain-language takeaway capturing the core idea to remember and apply immediately.
- **Bringing it together**: A brief synthesis that ties the chapter together and clarifies the practical shift you should make.
- **AI tools referenced**: A quick-reference list of the tools used so you can implement the ideas right away.

THE 42 WAYS AT A GLANCE

Here's a quick overview of the practical ways artificial intelligence can help simplify everyday life. You can read the book straight through or jump to the ideas that matter most to you.

Reducing Mental Load

1. AI as Your Daily Planner
2. Smart Scheduling and Conflict Resolution
3. Prioritization
4. Reminders and Task Tracking
5. Memory Backup
6. Auto-sorting, Summarizing, and Replies
7. Summarization Tools
8. Writing Kind but Firm Responses

Life Administration

9. Understanding and Working with Documents
10. Expense Tracking and Smart Budgeting
11. Meal Planning
12. Grocery Lists
13. Gift Ideas and Reminders
14. Travel Itineraries
15. Brainstorming
16. Curation for Leisure

Work and Creativity

Communication

Family and Home

Growth and Goals

CHAPTER 1

Reducing Mental Load: How to Regain Focus, Prioritize What Matters, and Stop Feeling Overwhelmed

Most days don't fall apart because of one big failure; rather, they unravel because of a hundred small decision-making moments that never stop coming.

- What should I work on first?
- Do I accept this meeting?
- Did I follow up on that message?
- Where did I put that document?

None of these questions is complicated, but answering them repeatedly drains focus faster than most people realize. By mid-morning, you may already feel behind, even though you have been busy the entire time. The issue is cognitive load, not effort.

This chapter shows how to use AI to reduce that load in practical, specific ways to spend less time sorting and more time doing. Let's make your day feel manageable again.

WAY #1: "WHAT SHOULD I DO FIRST?"

AI AS YOUR DAILY PLANNER

The first decision of the day shapes everything that follows. If you hesitate, you drift into low-effort activity like skimming emails or reorganizing tasks instead of executing them. Such hesitation is a form of "procrastination disguised as preparation," and it consumes cognitive bandwidth before real work can begin.

Most people fail to plan because their tasks are part of fragmented systems: Gmail, Slack, a notes app, a project manager, maybe a whiteboard, and whatever is just floating in your head. Your brain is an excellent problem-solver, but it is a poor sorting engine. When forced to reconcile these scattered sources every morning, you start the day in a state of decision fatigue.

Sorting and ranking your work—so you begin working with clarity instead of confusion—is one way AI can become useful in a practical way. The simplest approach uses ChatGPT. At the end of the workday, or first thing in the morning, paste your full task list and give it constraints.

AI Prompt:

```
Here are 12 tasks I need to complete this week. Rank them
by urgency and impact. Assume I have three hours of focused
work tomorrow morning. Identify the top three tasks I should
complete before noon.
```

Then paste your list.

ChatGPT will evaluate the tasks and produce a ranked response. It is not perfect, but it provides a structured starting point in seconds. Instead of staring at a pile of competing demands, you begin with a recommendation you can react to and refine.

If you use Notion for task management, Notion AI can perform this prioritization inside an existing database.

AI Prompt:

> Sort these tasks by due date and strategic importance. Highlight the tasks most aligned with my Q2 goals.

Because the data already lives in Notion, the prioritization reflects your existing structure rather than starting from scratch.

If you prefer deeper automation, Motion is a strong option. It automatically schedules tasks into your calendar based on deadlines and priority levels. When meetings are rescheduled, it handles the necessary changes. Instead of manually dragging calendar blocks around and recalculating what still fits, the system handles the adjustment.

Reclaim.ai works well if you rely heavily on Google Calendar. It protects focus time, schedules flexible tasks, and adjusts as your calendar shifts. You define priorities once, and the system continues defending your time in the background.

If your work flows through email and messaging tools, Sunsama can consolidate tasks from Gmail, Slack, Asana, and other systems into a daily planning view. You choose what goes into "Today," which forces intentional prioritization instead of reactive task switching.

For Microsoft users, integrating Copilot into Outlook and Planner can summarize outstanding tasks and surface commitments from recent emails and Teams threads. I've been using Copilot to do this for a year, and it works well. It reduces the time spent scanning for what matters and lowers the chance that something important stays buried in conversation history.

Even something as simple as Todoist with AI Assist can help. If your task list contains vague items like "Project proposal," the AI can rewrite them into specific actions such as "Draft first section of Q2 project proposal." Specific tasks are easier to prioritize and execute than abstract placeholders.

If you want a minimum-effort routine to start on today, keep it simple:

1. Maintain one master task list.
2. Open ChatGPT or Claude.
3. Paste your list and ask for the top three tasks for tomorrow morning.
4. Time-block those three into your calendar.

This takes five minutes, instead of the 20 minutes you would otherwise spend negotiating with yourself about where to begin.

In real life, this shifts the tone of your day. Upon opening your laptop, you already know your first move. That reduces procrastination disguised as preparation and makes it easier to resist distractions.

When a new request arrives at 9:30 a.m., you can compare it to your top three priorities. If it does not outrank them, it waits. You'll consistently complete high-impact work first, and your week will feel more intentional. Stress drops because you are no longer constantly second-guessing decisions.

Most task managers sort by due date, but "due soon" does not always mean "high impact." AI is useful here because it can evaluate the content of tasks against specific constraints. While a standard list treats a 10-minute administrative task and a two-hour strategic proposal with equal weight, AI can distinguish which moves the needle on your quarterly goals. It provides a recommendation based on more than chronology.

However, there are limits. AI can optimize a task list, but it cannot define values. If your commitments are unrealistic, the system will still attempt to schedule them. You must decide what truly belongs on the list. Also, minute-by-minute planning is not needed. Over-structuring a day can create more stress, not less. In most cases, three clearly defined priorities are enough to create momentum.

If you need AI to help adjust for fluctuations in energy levels, build that into the prompt. For example, tell the AI to schedule creative work in the morning and administrative tasks during the afternoon.

I recommend treating AI output as a draft. Review it, then adjust for energy level, personal commitments, and long-term goals. The tool helps you sort and compare; you make the final call. Used this way, AI does not run your day. What it does is remove the mental burden of deciding where to start. Instead of waking up and asking, "What should I do first?" you already have the answer.

Takeaway: When AI handles sorting and ranking, you preserve mental energy for doing the work that matters.

My Personal Experience

I do most of the work for my HomeTechHacker business on weekends. The work involves writing articles, creating social media posts, planning and writing books, accounting, strategic work, responding to emails, and many other things. As my business grew, the sheer number of tasks felt overwhelming. I'd often just react to the latest email or the most urgent social media post. Often, I would feel busy, but not as productive as I should be.

Then I started putting all my tasks into a Notion database. Additionally, I would track my quarterly, yearly, and five-year goals in Notion as well. Then, I started using Notion AI to suggest which tasks to prioritize each weekend. It is surprisingly efficient. I don't agree with everything it tells me, but most of the time I follow it, and I have felt much more productive.

WAY #2: "MY CALENDAR IS A MESS"

Smart Scheduling and Conflict Resolution

Look at next week's calendar. Meetings dominate, with small gaps scattered between them, and important projects appear out of nowhere. What does this say about your priorities? If your schedule feels chaotic, it is usually because it has expanded without deliberate design.

A calendar does not just show time. It shows commitments, and these erode your focus when they accumulate without review. Fragmentation becomes a hidden cost.

Ten 30-minute meetings scattered throughout a day can dismantle the ability to think clearly. Each meeting may seem reasonable in isolation, but the cumulative effect is constant context switching. You never settle into meaningful work before being pulled into something else. Even when there is "free time," it usually comes in short, unusable fragments, which is a pattern that drains momentum and increases stress.

AI tools are well-suited to this problem because scheduling is a constraint-based system, with fixed time blocks, deadlines, meeting durations, and availability windows. When those constraints are clearly defined, AI can optimize them far faster than you can manually.

If you use Google Calendar, using a tool like Reclaim.ai allows you to implement "flexible time blocking," which defines a priority once, and then the system defends that time as your calendar shifts throughout the week. It allows for the creation of flexible tasks that automatically schedule into open time blocks before a deadline.

Assign a duration and priority level, and the system places the task into your week, moving it if conflicts arise.

For example, you can create a task called "Draft marketing strategy," set it to two hours, and give it a deadline of Friday at 5:00 p.m. Reclaim.ai will find available space before that deadline. If someone books over it, the task is rescheduled automatically. You stop manually rearranging your week.

Motion takes a more automated approach. You enter tasks with deadlines and estimated durations, and it builds a dynamic schedule that continuously adjusts as meetings shift. It functions less like a calendar and more like a planning engine that recalculates throughout the day.

Microsoft users can rely on Copilot inside Outlook and Teams to surface meeting overload and highlight action items. It can also help identify recurring commitments that generate little value but consume a disproportionate amount of time.

You can also take a manual review approach with ChatGPT, copying an agenda from the "schedule" or "list" view of your calendar by taking a picture, printing to PDF, or exporting a calendar file and pasting it into the model with a prompt.

AI Prompt:

> Analyze this schedule. Identify where I have less than 30 minutes between meetings. Suggest how I could consolidate meetings and create two 90-minute focus blocks.

It will not move events, but will quickly reveal structural inefficiencies you may overlook.

A practical weekly workflow might look like this:

- Review your upcoming week every Friday afternoon,
- Identify your top two priorities,
- Schedule focused work before meetings fill the gaps, and
- Move or decline meetings that conflict with those priority blocks.

By doing this consistently, your week changes tone. The important work gets prime time, not the leftovers, and the resulting longer focus blocks reduce context switching after meetings. Projects move forward earlier, rather than being pushed to the end of the week.

AI can optimize your schedule, but it cannot manufacture time. If you consistently say yes to everything, no tool will solve that. Structural overload requires behavioral change, not just better automation. I'll cover "saying no" later in the chapter.

You also do not need to automate every minute. Some people prefer dynamic scheduling, while others prefer fixed blocks. Start by protecting high-value time and layer in automation if it adds value.

A good calendar has alignment with what is important, not just what is urgent. When time matches priorities, mental load drops, and your week becomes more productive and feels intentional.

> **Takeaway**: When a calendar reflects priorities instead of reactions, your week stops controlling you and starts working for you.

My Personal Experience

As my responsibilities as an IT director grew, my calendar slowly became a reflection of that growth. I realized I was constantly overwhelmed by meetings. Every day was packed with scattered 30-minute calls, each separated by short breaks that were too small to be useful. I was always switching contexts, jumping from one topic to another, and ending most days feeling busy but not like I had accomplished anything.

I finally turned to Copilot for help. I told it that my problem was not the total number of meetings, but how fragmented my time had become. Copilot reviewed my calendar patterns and highlighted how those short gaps were breaking my focus. It then helped me reorganize my schedule by grouping meetings together and intentionally creating protected 90-minute and two-hour blocks dedicated to deep work.

That change transformed how I work. Those longer blocks gave me the space to think strategically, solve complex problems, and actually move initiatives forward. My meetings became more purposeful, and days felt calmer and more productive. Using Copilot helped me take control of my time as my role continued to grow.

WAY #3: "I HAVE TOO MUCH ON MY PLATE"

PRIORITIZATION

The trap is thinking you have a time problem when you actually have a commitment problem, which usually accumulates quietly. When everything feels urgent, it is often because everything has been accepted: projects, favors, meetings, side initiatives, and quick requests. None of them seems unreasonable in isolation, which is why they slip through. Eventually, they create overload.

Most people do not consciously decide to overwhelm themselves. Their work builds gradually, one approval at a time, and because each decision feels small, the total weight remains invisible—until it becomes stressful. You rarely pause to evaluate the full picture. Instead, responding to what is in front of you, there is an assumption that you will adjust later. That "later" rarely arrives.

Another issue is structural. Most task lists are flat, meaning they do not show impact, cost, or strategic alignment. A minor administrative task can sit beside a revenue-driving initiative with equal visual weight. Without context, everything appears equally important, and that illusion makes trade-offs harder than they need to be. When urgency becomes the default sorting mechanism, strategy disappears.

AI can help you step back and triage at a higher level by forcing structured evaluation. Start with ChatGPT or Claude by pasting your current commitments, and not just today's tasks. Include recurring responsibilities, long-term projects, and personal obligations so the model has a full picture.

AI Prompt:

> Here is everything I am responsible for. Group these into high, medium, and low impact based on long-term goals. Identify which items could be deferred or delegated with minimal downside.

The output reframes your workload. Instead of reacting to deadlines, you begin seeing the relative impact of projects. That shift alone can change how you allocate your week.

If using Notion, create a simple database with fields for deadline, estimated effort, and strategic importance. Ask Notion AI to sort tasks by highest impact relative to effort. In many cases, you will discover that high-effort, low-impact work is consuming disproportionate time. Such visibility makes decisions easier because the trade-offs become explicit.

For managers and business owners, tools like ClickUp AI or Asana Intelligence can summarize workload distribution across projects and find "hidden" bottlenecks. Instead of scrolling through hundreds of tasks, you can ask the AI: "Which team member is the primary blocker for our Q2 goals?" This might surface four separate projects currently stalled because all are awaiting a single person's approval. This allows you to reassign those approvals or clear the logjam immediately, the kind of insight often buried in a standard, flat task list.

Consider this practical example. A product launch, weekly report, mentoring responsibilities, and a volunteer commitment are being juggled. ChatGPT may highlight that the product launch drives measurable revenue, while the reporting format could be simplified or partially automated. That distinction does not eliminate work, but it clarifies leverage. A few structural adjustments could free several hours per week without reducing any meaningful output.

Another approach is to simply eliminate low-value tasks. Instead of asking AI to help you fit everything in, ask it to help you cut things out.

AI Prompt:

> ```
> Identify the bottom 20% of these tasks in terms of strategic
> value. Which of these could be deferred, delegated, or
> eliminated entirely with minimal downside?
> ```

This reframes the goal from "doing more" to "choosing better." It exposes the often invisible trade-offs, making it easier to stop doing work that no longer serves your primary objectives.

Unless you provide context, AI will not understand office politics, the nuances of long-term relationships, or personal values. If the input is shallow, recommendations will be shallow as well. Also, do not use AI as an excuse to shirk responsibilities without thinking things through. Use it as a structured partner that sharpens judgment rather than replacing it.

If your plate feels full, the answer is rarely to work faster; it is to be found through different decisions.

> **Takeaway**: When everything feels urgent, use AI to expose trade-offs, allowing you to choose intentionally rather than reacting automatically.

WAY #4: "I FORGET FOLLOW-UPS"

Reminders and Task Tracking

A missed follow-up is a silent hit to your professional reputation. Most people fail to follow through, not because they are lazy, but because they rely on memory. Without a system to convert a conversation into a tracked action, there is a commitment only as long as your short-term memory allows.

AI-powered tools help escape this problem by converting unstructured communication into structured reminders. After a meeting, paste key points into the model.

AI Prompt:

```
Extract clear follow-up actions from this meeting summary and
format them as task list items with suggested deadlines.
```

Immediately copy the items into Todoist, Microsoft To Do, or Notion. The key is not the tool but the consistent, quick conversion of conversations into tasks.

More automated options exist. Superhuman and Shortwave can detect action items in email threads and suggest reminders. Microsoft Copilot can surface commitments made in Teams chats and Outlook conversations, reducing the chance they disappear into history.

A simple workflow works well. After a meeting or important email, paste the contents into ChatGPT, extract action items with suggested due dates, and move them into your task manager before closing the conversation. This is a habit that ensures commitments become visible tasks instead of mental notes.

AI Prompt:

> ```
> From the email below, identify any commitments I made and
> suggest a follow-up date for each.
> ```

This reduces interpretation time and forces clarity. Instead of scanning for implied obligations, a clean list to confirm and execute was produced.

Another option is to capture follow-ups automatically during the meeting itself. Many video conferencing tools now include AI note-taking features that can record conversations, generate summaries, and extract action items.

Tools like Zoom AI Companion, Microsoft Teams Copilot, and Google Meet AI can identify tasks discussed during a meeting and then present them as follow-ups. Instead of relying on memory or notes, a structured note is produced, including a summary of suggested action items immediately after the conversation ends.

This increases reliability. You will stop apologizing for missed follow-ups and start closing loops consistently. Relationships improve because others can depend on you, and you reduce background stress because unfinished commitments are no longer floating in your head.

There are boundaries to consider. AI tools that scan email may raise privacy concerns, especially in regulated industries. Review your company's policies before enabling automated analysis. I also suggest adding a verification step where you double-check for drift, such as AI turning a "maybe" or suggestion into a hard commitment.

Do not automate blindly. Capture commitments intentionally, confirm they are real, and ensure they align with your priorities before adding them to your system.

> **Takeaway**: Capture follow-ups immediately with AI support so that your memory is not the weak link in your reliability.

WAY #5: "I KEEP FORGETTING EVERYTHING"

MEMORY BACKUP

Your brain is optimized for thinking, not storage; however, many of us treat it like a digital archive. Password hints, half-formed ideas, contractor estimates, book recommendations. Everything gets pushed into mental memory, with the expectation that it will all remain accessible. That overwhelms us.

When information lives only in your head, retrieval becomes unpredictable.

- You know you saw the document somewhere.
- You remember discussing a detail with someone.
- You recall saving a link, but cannot find it.

That kind of mental load compounds, wasting energy that is better spent on analysis or decision-making.

AI functions as a “second brain” because it understands the meaning of your notes, not just the keywords. In a traditional notes app, searching for “expenses” will only return notes containing that exact word. With an AI-powered system like Notion AI or Evernote’s AI search, you can ask, “How much did we spend on the kitchen project?” Even if the notes only mention “contractor invoices” and “tile receipts,” the AI understands those are expenses and retrieves the relevant data. This “semantic search” means less time spent meticulously tagging notes and more time actually using them.

To do this effectively, use a “dump and distill” workflow. Once a week, paste your scattered notes, links, and half-formed ideas into ChatGPT or Claude.

AI Prompt:

```
Categorize these notes by project. Summarize the key data points
for each and list any hidden deadlines I may have mentioned.
Format this as a clean reference guide.
```

This transforms a pile of digital clutter into an actionable resource. It allows you to stop digging through threads and start focusing on the actual application of the information.

Another option is to use tools like Notion AI, Evernote with AI search, or Obsidian with AI plugins to capture and retrieve information efficiently. These tools can distill information into more usable and actionable content. I recommend just using one of these tools at a time to start.

As an example, imagine planning a home renovation. Contractor quotes, product links, measurements, and design ideas are scattered across apps and emails. Instead of searching multiple locations, they are consolidated into one document, and a structured summary with action items is generated. Retrieval becomes predictable rather than frustrating.

The result is faster access and reduced mental clutter. You stop digging through threads or scrolling across multiple platforms, as the information has a defined home. Your mind is free to focus on thinking rather than storing.

Most general AI tools do not retain memory across sessions unless specifically configured to do so. Do not assume ChatGPT will remember something from last month unless it is saved externally. Also, be cautious with sensitive data, and avoid storing passwords or confidential information in unsecured systems.

Takeaway: Use AI to organize and retrieve information so that your mind can focus on thinking instead of storing.

My Personal Experience

A couple of years ago, I decided to get serious about my fitness. I wanted to lose weight, run races, get stronger, and achieve other fitness goals. I use several different tech tools to track my weight, workout heart rate, weights lifted, sleep recovery, body fat percentage, fitness goals, etc. Keeping track of all these things and using the data effectively was impossible.

I started logging information from my workouts and listing my goals in a ChatGPT "Fitness" project. I upload data files from my runs, provide updates on daily exercises and weight, and include goals and workout plans. With that information, it keeps track of how much I'm running, how many miles I've put on my running shoes (so I know when to replace them), and provides feedback on progress and advice on training plans. I can even ask about old workouts—how I felt, what I've improved, what has gotten worse—and I don't have to try to remember all of these things.

I'm in the best shape of my life, and next to all the work I've put in, the advice and strategic, data-informed planning I've gotten from ChatGPT is the second biggest reason why.

WAY #6: "I'M DROWNING IN EMAILS"

AUTO-SORTING, SUMMARIZING, AND REPLIES

Email overload is a decision-making problem, not a volume problem. Every message in the inbox is a micro-decision waiting to be made: respond, archive, or defer. When these decisions accumulate, focus shifts from producing meaningful work to merely managing inputs.

AI tools reduce the mental burden of sorting and drafting by compressing information before engaging with it. Gmail's "Help Me Write" and summary features condense long threads. Microsoft Copilot in Outlook can summarize email chains and draft responses. Superhuman and Shortwave categorize and prioritize messages automatically, helping you focus on what matters.

You can also use ChatGPT directly.

AI Prompt:

> `Summarize this email thread in five bullet points and identify the action I need to take.`

Paste the thread after the prompt. Instead of parsing every line, you can review a structured summary and decide how to respond.

For faster triage, combine filters with AI summaries. Have emails from key clients forwarded into a priority folder, then summarize only that folder each morning, and draft replies with AI assistance before editing for tone. That approach narrows attention to high-impact messages rather than letting the inbox dictate your day.

This practice reduces "inbox anxiety." Instead of reading every word of a 15-message exchange, you review a concise overview and act. Response times will improve, decisions will feel clearer, and email becomes a tool instead of a drain.

However, AI-generated replies must be reviewed carefully. Tone errors or subtle inaccuracies can damage trust, especially in professional settings. Be cautious when sharing confidential threads with external systems and, as always, confirm compliance requirements before enabling automation.

> **Takeaway**: Let AI compress the noise in your inbox to focus on messages that require attention.

WAY #7: "I DON'T HAVE TIME TO RESEARCH THIS"

Summarization Tools

Information is not lacking; time to process it is. Modern decision-making often means long PDFs, articles, and reviews. The bottleneck is filtering the signal from the noise. When the volume of raw material exceeds your capacity to read it, decisions stall.

AI summarization tools address that bottleneck directly by condensing large amounts of content into structured insight. Perplexity AI combines search with summarized answers and citations, making it strong at comparison research.

AI Prompt:

> Compare the top three CRM platforms for small businesses under 50 employees. Summarize pros and cons with pricing.

Claude handles long-document summaries well, for example, lengthy PDFs.

AI Prompt:

> Summarize this in 10 bullet points and identify key risks.

ChatGPT can generate executive summaries.

AI Prompt:

> Create a one-page summary of this 20-page report, focusing on financial implications.

In practice, this changes how research is approached. Before purchasing new business software, summarize vendor documentation and compare features side by side. Instead of scanning dozens of pages manually, evaluate structured output and focus on the trade-offs. While controlling and guiding the analysis, reading time is reduced. This results in faster, more informed decision-making. Less time is spent absorbing raw material, more evaluating implications. The work becomes decision-focused rather than document-focused.

AI summaries can miss nuance or misinterpret data, especially in technical or high-stakes contexts. Always verify critical information directly from primary sources. Do not rely solely on summarized output for legal, medical, or financial decisions.

> **Takeaway**: Let AI distill large amounts of information to decide efficiently without drowning in details.

WAY #8: "I'M BAD AT SAYING NO"

Writing Kind but Firm Responses

Saying "yes" feels easier in the moment because the discomfort of a "no" is immediate. However, the consequences of "yes" are cumulative. Every minor agreement dilutes priorities until your schedule is no longer your own. While wanting to appear helpful, saying yes to everything makes it impossible to be truly effective at anything.

The challenge is rarely a lack of will; rather, it is a lack of ready language. When a request arrives, the pressure to respond quickly often leads to a soft "yes" that is later regretted. AI acts as a communication buffer, helping to draft clear, respectful boundaries so you can express a firm "no" without the typical internal negotiation or guilt.

AI can assist in communicating "no" decisions cleanly. ChatGPT and Claude are particularly useful when tone matters, because you can provide context and refine the response until it feels right.

AI Prompt:

> I have been invited to a weekly meeting that overlaps with my priority project work. I want to decline respectfully and offer an alternative way to stay informed. Draft a concise response.

The initial output provides structure. From there, you can tighten or adjust tone with follow-up prompts such as, "Make this firmer and shorter," "Reduce apologetic language," or "Keep it professional but not cold." Instead of wrestling with wording for ten minutes, you refine a draft that already communicates the core message.

You can create a reusable framework.

AI Prompt:

> Write a short email that declines a request due to current capacity, affirms the relationship, and suggests a next step.

That template works across many scenarios. For Slack or Teams, where brevity matters, use a slightly different approach.

AI Prompt:

```
Draft a two-sentence response declining a request to take on
additional work this week because I am at capacity.
```

AI can also help before you respond.

AI Prompt:

```
Based on my stated priorities of X and Y, does this align
with my current goals?
```

Seeing the mismatch articulated clearly often provides the confidence to follow through. Sometimes the issue is not language but clarity about trade-offs.

The benefit of doing this consistently is fewer ill-timed commitments, cleaner communication, and reduced resentment. When declining clearly and respectfully, most people adjust quickly. Boundaries become predictable instead of emotional, and you stop carrying frustration from agreements you never wanted to make in the first place.

AI-generated responses can sound generic if accepted without editing, which can weaken trust. You should also avoid sharing sensitive workplace dynamics (e.g., internal budget figures, project code names, sensitive research topics) in public AI tools. In some situations, especially when the stakes are high or relationships are strained, a live conversation is better than a message.

AI shortens the path from hesitation to clarity and lowers the friction of expressing what you already know to be true. It does not replace judgement. If your schedule feels overloaded, learning to say no is one of the highest-leverage skills to develop. AI simply makes that skill easier to practice consistently.

> **Takeaway**: Use AI to draft messages that build clear, respectful boundaries that protect your time without damaging relationships.

BRINGING IT TOGETHER

When your brain is constantly juggling tasks, schedules, and decisions, it is easy to feel behind. The tools in this chapter will not eliminate your workload, but they will remove unnecessary friction. Clear priorities, structured calendars, reliable follow-ups, cleaner inboxes, and firmer boundaries reduce the noise competing for your attention. Even small adjustments, such as defining three top tasks or protecting a focus block of time, can change how your week feels. This will lead to fewer unnecessary decisions and more control over your time.

AI TOOLS REFERENCED

Start simple. You can reduce mental load with just ChatGPT and one scheduling tool, then expand when ready.

- **ChatGPT (OpenAI)**: Used for task prioritization, extracting follow-ups, drafting responses, analyzing schedules, summarizing emails, and synthesizing research. Best as a flexible thinking and drafting partner.

- **Claude (Anthropic)**: Strong for long-document summaries and structured analysis. Useful for reviewing commitments, summarizing reports, and drafting tone-sensitive communication.
- **Notion AI**: Integrated inside Notion databases and notes. Helps organize task lists, group priorities, summarize notes, and build a structured "second brain."
- **Motion**: Automated task scheduling tool that builds and dynamically adjusts your calendar based on deadlines and priorities. Best for users who want high automation.
- **Reclaim.ai**: Integrates with Google Calendar to automatically schedule flexible tasks and protect focus time. A balanced option between manual control and automation.
- **Sunsama**: Consolidates tasks from Gmail, Slack, Asana, and other tools into a daily planning workflow that forces intentional prioritization.
- **Zoom AI Companion**: Built-in AI features within Zoom that can summarize meetings, generate notes, and identify follow-up tasks. It helps ensure that important action items discussed during meetings are captured without relying on manual note-taking.
- **Microsoft Copilot (Outlook / Teams / Planner)**: Surface commitments from email and chat threads, summarize meetings, and assist with task extraction for Microsoft ecosystem users.
- **Google Meet AI**: AI-powered features in Google Meet that can generate meeting summaries, capture key points, and suggest follow-up actions. Useful for keeping track of decisions and commitments made during conversations, especially when meetings move quickly.
- **Todoist (with AI Assist)**: A task manager that helps clarify and rewrite vague tasks into actionable steps for easier prioritization.

- **Superhuman / Shortwave**: AI-enhanced email clients that categorize, summarize, and draft responses to reduce inbox overload.
- **Perplexity AI**: AI-powered research assistant that combines search results with summarized answers and citations for faster comparison and evaluation.
- **Evernote or Obsidian (with AI plugins)**: Knowledge management tools that support structured note storage, search, and AI-assisted organization.

CHAPTER 2

Life Administration: Building Systems for Everyday Life

Work is not the only place where decisions pile up. Outside of meetings and deadlines, there is another layer of life administration that rarely gets acknowledged:

- Meals to plan,
- Groceries to buy,
- Documents to locate,
- Birthdays to remember,
- Trips to organize, and
- Bills to track.

None of these tasks is especially difficult; however, they demand steady attention. Over time, the background demand can feel heavier than the tasks themselves.

This chapter focuses on simplifying those everyday responsibilities. How AI can help generate plans quickly, organize recurring tasks, structure information, and reduce the small inefficiencies that drain attention throughout the week. The aim is to make routine logistics easier to manage, so energy is reserved for decisions that deserve it. Life feels easier when your daily administration runs smoothly.

WAY #9: "WHERE DID I PUT THAT DOC?"

Understanding and Working with Documents

Searching for a document usually starts the same way. You know it exists, but you cannot remember where you saved it or what you called it. You check your downloads folder, email attachments, cloud storage, or desktop. Eventually you find the file, but a second problem appears.

What exactly does it say?

Many documents are difficult to interpret quickly. Insurance policies, contracts, invoices, travel confirmations, meeting notes, and long PDFs often contain the information you need buried inside pages of text. Finding the file is only half the challenge. Understanding and using the information inside it is often the bigger problem.

Traditional search depends heavily on filenames and exact keywords. AI-powered search works differently. It relies on meaning.

With Google Drive AI search or Microsoft Copilot, you can search using natural language instead of remembering exact titles. Rather than typing a filename, you might search:

AI Prompt:

```
Find the invoice for the plumber who repaired the kitchen
sink last summer.
```

The system can understand that "plumber" refers to a contractor and that "invoice" may appear as a receipt, estimate, or billing document. Instead of relying entirely on memory, you search based on context.

That alone saves time, but AI becomes even more useful after the document is located.

Suppose you find a 40-page insurance policy or legal agreement. Most people do not want to reread the entire document just to answer a single question. Uploading the file into ChatGPT or Claude can dramatically shorten that process.

AI Prompt:

```
Summarize this document and extract key dates, obligations,
and decisions.
```

In seconds, AI can produce a structured summary that highlights the information most likely to matter.

This approach works especially well for documents that are information-heavy or difficult to interpret. Examples include:

- insurance policies

- contracts and agreements
- medical paperwork
- meeting notes and transcripts
- travel confirmations
- lengthy PDFs or reports

AI can also help explain information that feels technical or confusing.

AI Prompt:

> ```
> Explain this document in plain language and highlight anything I
> should pay attention to.
> ```

Instead of struggling through unfamiliar terminology, you receive a simplified explanation that makes the document easier to understand and act upon.

Another powerful capability is comparison. Sometimes the challenge is not understanding one document but understanding differences between multiple versions.

AI Prompt:

> ```
> Compare these two insurance policies and summarize the major
> differences in coverage, exclusions, and deductibles.
> ```

This kind of analysis would normally require careful side-by-side reading. AI can surface important differences quickly and provide a clearer starting point for decision-making.

The broader shift is moving from document storage to document usability. Instead of asking, "Where did I save that?" or "Do I have time to read this?" AI allows you to focus on the information itself. Documents become easier to search, understand, and use.

AI search depends on readable content. Documents scanned only as images may not work as well unless optical character recognition (OCR) has converted them into searchable text. Privacy also matters. Uploading confidential documents into public AI systems may violate workplace or legal requirements. Review any policies that apply before processing sensitive files.

AI should support understanding, not replace judgment. Summaries and explanations are starting points, not substitutes for reviewing important details yourself.

Takeaway: Use AI to find, understand, and work with documents so that information becomes usable instead of overwhelming.

WAY #10: "MY BUDGET'S A MESS"

Expense Tracking and Smart Budgeting

Stress about money is rooted in uncertainty as much, if not more, than overspending. You may know your major expenses, but small transactions like subscriptions, coffee, and Amazon impulse buys accumulate quietly. Without consistent categorization, spending patterns remain invisible. Checking your bank balance does not reveal trends; it only shows the current state.

Automated categorization combined with regular review produces clarity. Tools like Monarch Money, YNAB, and Copilot Money learn transaction patterns. They make it easy to find trends.

No matter which budgeting tool is being used, I recommend reviewing uncategorized or miscategorized transactions weekly. After a month or two, the system can be taught your habits and then improve accuracy. Once the data is clean, AI becomes useful for analysis rather than tracking.

You can use an AI budgeting assistant or export a month or quarter of categorized spending and paste the summary into ChatGPT.

AI Prompt:

```
Analyze this spending breakdown. Identify categories that have
grown over the past three months and suggest areas where small
adjustments could reduce total monthly spending by 10 percent.
```

This moves budgeting from restriction to optimization. You're identifying trends instead of guessing where to cut.

AI is most effective when used for "what-if" planning, which is difficult to calculate mentally. Once you have a month of categorized spending, the AI assistant can be asked questions.

AI Prompt:

> ```
> If I reduce my dining out budget by 20% and cancel my three
> least-used subscriptions, how much faster will I reach my $5,000
> savings goal?
> ```

Instead of looking at a static list of past mistakes, you are using data to predict future outcomes. This transforms budgeting from a source of guilt into a strategy tool that allows you to see, in seconds, the long-term impact of small daily changes.

A 20-minute monthly review—analyzing trends, finding unnecessary subscriptions, and ensuring accurate expense categorization—is more effective than constant anxiety about individual purchases. The key is having budget and spending data that AI can analyze.

For irregular expenses, such as travel or home repairs, use sinking funds (money set aside regularly for a planned future expense) inside Monarch Money or similar tools. Allocate a fixed amount monthly so large expenses do not feel like emergencies. AI can help determine appropriate contributions.

AI Prompt:

> ```
> Given an annual home maintenance estimate of $3,000, calculate a
> monthly allocation.
> ```

Remember, financial AI analysis depends entirely on clean data. If transactions are miscategorized or accounts are missing, the insights will be flawed. Privacy also matters. Only export or analyze data in tools that meet security standards.

Takeaway: Use AI-supported tracking and analysis to replace guesswork with visibility and structured review.

My Personal Experience

I have been using Monarch Money for years. It has made tracking and analyzing spending much easier. I used to track everything in a spreadsheet, and as our finances became more complicated, this became too time-consuming. Instead, I would just glance at my balances and occasionally check my credit card statements for fraud and mistakes.

Monarch Money automatically aggregates all that tracking into one place. Weekly, I review all charges and track spending against our budget with ready-made reports. I also use its AI assistant to analyze spending trends, suggest improvements, and even give thoughts on how to increase my net worth. It has been an invaluable tool in optimizing money and planning for the future.

WAY #11: "WHAT'S FOR DINNER?"

MEAL PLANNING

Dinner doesn't become stressful at 6:00 p.m. It does so at 3:00 p.m., when realizing you haven't decided what to cook. By evening, tired and hungry, you are negotiating with whatever happens to be in the refrigerator. That combination almost guarantees a rushed or repetitive choice.

Treating dinner as a daily decision, which consumes daily energy, is the real issue. When choosing is postponed until you are depleted, you default to whatever is fastest or familiar. That may work occasionally, but over time, it increases spending, food waste, and monotony.

A better approach is shifting from improvisation to weekly planning. Decide once, execute multiple times. Meal planning fits AI well because it involves clear constraints: budget, schedule, dietary preferences, and ingredients on hand. When the constraints are defined, structured options are easy to come by.

Start by giving ChatGPT real parameters.

AI Prompt:

```
Create a 5-day dinner plan for two adults and two kids.
Meals must take under 30 minutes, reuse ingredients where
possible, and keep total grocery cost under $100.
```

That specificity produces usable output. If the plan includes unnecessary ingredients, refine.

AI Prompts:

```
Reduce ingredient variety and reuse proteins across multiple meals.

Add one slow cooker option for a busy day.

Swap one meal for something vegetarian.
```

Here, the strength of AI is iteration speed. Instead of scrolling recipe sites for 40 minutes, you narrow toward a workable plan in five.

Do not treat meal planning as a weekly creative exercise. Instead, build a "Recipe Vault" in a simple document or notes app. When the AI generates a plan you enjoy, paste the prompt and the result into the vault. Eventually, you can stop asking the AI for new ideas.

AI Prompt:

```
Reorganize Week 4 of my Recipe Vault into a new grocery list based on current sales.
```

A limitation of this approach is that the AI does not know what's in the pantry unless you tell it. If vague constraints are provided, you will receive vague suggestions. Review every plan for practicality before committing to it, especially if allergies or strict dietary goals are involved.

> **Takeaway**: Choose a weekly dinner plan with AI support once a week and stop renegotiating it every evening.

WAY #12: "I HATE GROCERY SHOPPING"

Grocery Lists

An efficient grocery shopping trip is really an exercise in logistics, and very frustrating when that structure is lacking. When working from a partial list and filling the rest from memory, you end up doubling back through aisles and grabbing items you did not plan to buy. The pattern repeats itself when nothing structural changes.

The mental load of grocery shopping usually stems from separation. Recipes live in one place, while the pantry inventory lives somewhere else, and the shopping list forms in your head. Without consolidating these, you rely on recall instead of design.

AI helps most when it consolidates inputs instantly. After generating a weekly meal plan, paste it into ChatGPT.

AI Prompt:

```
Create a consolidated grocery list grouped by store sections
and remove duplicate ingredients.
```

Grouping by produce, protein, dairy, and dry goods reduces backtracking in the store. It turns the trip into a linear path instead of a maze.

If using AnyList, take advantage of its aisle mapping so the list follows the store's layout. If using Apple Reminders, create sections manually so the list mirrors the path you take in the store.

For households that monitor spending closely, another layer can be added via ChatGPT.

AI Prompt:

```
Estimate the total cost of this list using average U.S. grocery
prices and flag the highest-cost items.
```

The estimate will not be perfect, but it will highlight where adjustments may be needed before going to the store.

Use the AI to audit your budget, then generate a shopping list.

AI Prompt:

```
Identify the three most expensive items on this list and
suggest a lower-cost alternative for each.
```

This forces a conscious decision about high-cost items before you are standing in the aisle under the influence of marketing.

Keep in mind that AI price estimates vary by region and should not be treated as exact. Also, if you do not keep track of what you already have at home, you may end up buying duplicates. This system works best when inputs stay current and accurate.

> **Takeaway**: Use AI to generate structured grocery lists so that shopping is efficient rather than reactive.

WAY #13: "I FORGOT A BIRTHDAY AGAIN"

Gift Ideas and Reminders

Most people remember birthdays; they remember them the week before, sometimes even the day before. However, they have forgotten to act early enough to do something thoughtful. By the time the calendar reminder fires on the morning of the event, the only options left are rushed messages or overnight shipping.

Giving birthday gifts work better when treated as small projects. There is planning, purchasing, possibly shipping, and sometimes coordination. Compressing all of that into the event day creates avoidable pressure.

One way to address this is to create recurring annual events with two reminders. In Google Calendar or Apple Calendar, create recurring annual events and set two reminders: one two weeks in advance and one three days before. The first reminder is the planning window. The second is a deadline to execute. That will eliminate most last-minute scrambling.

Centralize gift ideation by creating a dedicated "Gift Ideas" note. Use a lightweight system such as Notion, Apple Notes, or even a structured Google Doc to maintain a running list of gift ideas tied to specific people. When someone casually mentions a hobby, a product they like, or something they need, add it immediately. If using Notion AI, tag entries by category and ask it to summarize past gift ideas before a birthday arrives.

AI becomes useful at the ideation and refinement stage, not just the brainstorming stage. When the two-week reminder triggers, combine stored notes with structured prompts.

AI Prompt:

> Based on these interests and previous gifts, suggest three thoughtful gift ideas under $100 that feel personal and not repetitive.

This approach produces stronger ideas as it's grounded in context, without which AI suggestions tend to be generic.

You can also use built-in recommendation engines more deliberately. Amazon and Etsy both use AI-driven personalization. Instead of browsing randomly, search within specific categories tied to the recipient's interests. Pair that with Perplexity to quickly compare product quality or read summarized reviews.

AI Prompt:

```
Summarize the pros and cons of this product based on recent reviews.
```

This step helps avoid impulse purchases that look good but disappoint in practice.

For people who struggle with remembering dates, automation helps. Services like Google Assistant, Apple Reminders, or Microsoft To Do can be set to recurring annual reminders tied to contacts. Some CRM-style apps, such as Clay or even simple contact apps with AI features, can surface upcoming birthdays automatically.

An important benefit of systematizing birthdays is consistency. When reminders trigger early, and gift ideas are stored year-round, pressure lessens. You're no longer relying on recall or last-minute creativity. The system carries the load, and you step in only to make final decisions.

AI-generated ideas will only be as thoughtful as the input provided. A vague description produces predictable results. Also, if care isn't taken, over-automation can make gestures feel transactional.

> **Takeaway**: Build a two-stage reminder system and use AI-assisted ideation so that important dates never depend on last-minute memory.

My Personal Experience

I'm usually good at remembering birthdays, but not perfect. Over the years, there have been a few times when I have forgotten the exact date, or just remembered too late to do anything thoughtful. To stop this from happening, I created a specific Google Calendar for birthdays. I have it default to reminding me two weeks in advance so I can plan something or mail a card if necessary. It also gives me a reminder the day of, so I can call or text. I even store the year of their birth in the calendar item, so I'll always know how old they are in case it's a milestone birthday.

WAY #14: "I CAN'T PLAN THIS TRIP RIGHT NOW"

Travel Itineraries

Trip planning often feels overwhelming. There are so many different decisions that need to be coordinated, including flights, lodging, transportation, activities, and timing. Even a short trip introduces dozens of decisions. When all of them compete, planning hits analysis paralysis.

This leads to fragmentation, which often tires our minds. Flights get researched one day, hotels another. Notes scatter across tabs and inboxes. As departure approaches, preparation feels heavier than the trip itself. AI is helpful here because travel planning requires filtering and sequencing.

Start with destination clarity. Before opening ten browser tabs to research everything at once, use Perplexity or ChatGPT to narrow options.

AI Prompt:

```
Plan a 4-day trip to Denver in September for two adults.
Focus on outdoor activities, good coffee, and moderate budget
accommodations. Include a rough daily outline.
```

This will produce a scaffold, not a final plan. You now have a structured starting point instead of a blank page.

For flights, use Google Flights or Kayak with flexible date search enabled. Google Flights uses predictive pricing models and fare tracking. Once viable options are identified, set price alerts rather than checking manually. That removes daily monitoring and shifts the process to notification-based.

For accommodations, use AI-assisted filtering rather than browsing endlessly. On Airbnb and Booking.com, refine filters aggressively before reading reviews. Then use a summarization tool to compress reviews.

AI Prompt:

```
Summarize the most common complaints in recent reviews of
this property.
```

This prevents overreaction to isolated negative comments.

Once major bookings are complete, shift to the itinerary's structure. Use ChatGPT to sequence activities logically.

AI Prompt:

> Organize these attractions into a 3-day itinerary grouped by geographic proximity and estimated travel time.

Paste your list. This will prevent backtracking across a city and save hours.

For organization, store confirmations in one place; Google Drive, Apple Notes, or a dedicated travel folder in Notion works well. Then paste confirmation emails into ChatGPT.

AI Prompt:

> Extract key reservation details and format as a clean itinerary with dates, addresses, and contact numbers.

This produces a single reference document instead of scattered emails.

AI-supported travel planning lightens the load of decision-making. Instead of making 50 isolated choices, make structured batches of decisions instead. Plan flights first, lodging second, and group activities by day. That sequencing reduces mental overhead and shortens planning time.

Use AI to reduce reading volume, but do not use it to eliminate final verification. AI models can occasionally "hallucinate" specific details, such as a restaurant's opening hours, a current ticket price, or the specific availability of a niche product. Treat AI-generated research as a high-quality draft that requires a quick manual check before booking a flight or committing to a purchase.

> **Takeaway**: Use AI to structure and sequence travel decisions so that planning becomes organized instead of overwhelming.

WAY #15: "I NEED A SIDE HUSTLE IDEA"

Brainstorming

When someone says they need a side hustle, what they often mean is that they need a viable idea. The internet offers endless lists of generic options: dropshipping, affiliate marketing, print-on-demand, coaching, and flipping products. The volume of suggestions creates noise rather than clarity.

Generating options is easy. Evaluating them against your time, skills, and market reality is harder. That evaluation step determines whether an idea becomes viable.

AI is useful for structured brainstorming when constraints are provided. Start with ChatGPT and frame the request carefully.

AI Prompt:

> Generate 10 side hustle ideas that leverage my background in IT leadership, require less than 10 hours per week to start, have low upfront cost, and can scale to $2,000 per month.

Specific parameters force the model to narrow toward realistic paths. Then refine.

AI Prompts:

> Rank these ideas by startup complexity.
> Identify which ideas have recurring revenue potential.
> Flag ideas that rely heavily on paid advertising.

This second layer is where real value appears. You are no longer collecting ideas; you are evaluating them.

Don't spend weeks on "market validation." Instead, use AI to find out if people actually want to buy what you are offering. Perplexity is a great tool for this.

AI Prompt:

> Summarize current demand and competition for offering home network consulting services in mid-sized U.S. cities.

This gives directional insight without hours of manual searching.

If wanting structured scoring, create a simple evaluation grid in Notion or a spreadsheet with criteria such as startup cost, time requirement, skill fit, market demand, and scalability. Paste candidate ideas into ChatGPT.

AI Prompt:

> Score these side hustle ideas from 1-5 on each criterion and justify the score briefly.

The output is not definitive, but it will help highlight trade-offs clearly. Instead of drifting between possibilities for weeks, compress exploration into structured sessions. Instead of asking "What could I do?" ask "Which of these fits my constraints and strengths?" That reduces indecision.

Generating options is just the first step. The real mental load of a side hustle is the administrative setup: tracking hours, managing invoices, and evaluating market viability. AI is most useful when it acts as a business analyst for specific constraints.

Start by using a "Feasibility Prompt" to pressure-test an idea before committing.

AI Prompt:

> ```
> I am considering starting a home network consulting business.
> List five potential 'hidden' administrative tasks this will
> require and estimate how many hours a week I will spend on
> billing versus actual work.
> ```

Once an idea passes that test, use AI to create initial operating documents.

AI Prompt:

> ```
> Draft a simple services agreement and a professional invoice
> template for a freelance IT consultant.
> ```

This moves you from the "dreaming" phase to the "operating" phase without having to start from a blank page.

AI cannot assess motivation, risk tolerance, or long-term commitment. It also cannot guarantee market success. Treat its output as directional analysis, not validation. Real validation requires testing, conversations, and small experiments. Side hustles succeed when they align with skills and available capacity.

> **Takeaway**: Use AI to generate and systematically evaluate side hustle ideas so that choices are based on fit and feasibility.

My Personal Experience

In addition to writing books, I run HomeTechHacker.com, a site dedicated to helping simplify tech so everyone can easily use it to their advantage. I am always coming up with ideas, though admittedly, some of them are bad. Sometimes they take too much time and money, don't add much value, or nobody wants them. Now, I use ChatGPT to filter my ideas based on things like effort, value, and profitability. This even helps me figure out topics for articles and books.

WAY #16: "I NEVER KNOW WHAT TO WATCH/READ"

CURATION FOR LEISURE

You open Netflix or Amazon Prime with the intention of unwinding. Twenty minutes later, you're still browsing. The same thing happens with books. You search, skim reviews, add titles to a list, and close the app without choosing anything. You have too many choices and too much information.

Most platforms already use AI to recommend content, but passive recommendations rarely work well because the inputs are weak. If you randomly watch different genres or never rate content, the algorithm cannot learn your preferences accurately.

On streaming platforms like Netflix, Hulu, or Prime Video, consistently rate what you watch. Even a simple thumbs up or down improves the recommendation model. The system adjusts faster when provided feedback. You can also use tools like Likewise or Taste.io, which are built specifically around preference modeling and AI-driven recommendations. These platforms ask structured questions about tone, genre, pacing, and themes, which significantly improves suggestion quality.

For books, Goodreads and StoryGraph use machine learning to refine recommendations based on ratings and reading patterns. StoryGraph, in particular, allows filtering by mood, pacing, and length. Instead of searching broadly, you can narrow things down.

AI Prompt:

> Fiction, medium pace, character-driven, under 350 pages.

If wanting deeper curation, move beyond genre and ask the AI for specific "mood matches." Instead of searching for "Science Fiction," provide the AI with a list of books or movies you already love and ask for a specific aesthetic connection.

AI Prompt:

> I want a book that feels like the movie *Inception*—something with a fast-paced, psychological puzzle at its center that deals with layers of reality.

This approach forces the AI to factor the structural and emotional DNA of a story rather than just the library category. It turns a generic search into a targeted recommendation that is much more likely to result in a satisfying choice.

Perplexity can help when you want curated lists tied to current trends.

AI Prompt:

```
List the most critically acclaimed sci-fi novels released in
the past two years with average ratings above 4.0.
```

This pulls in recent data and summarizes quickly.

When preferences are clear and recommendation systems are trained, decision time drops from 20 minutes to two. You move from endless browsing to quick and accurate selection.

There is one practical boundary to watch out for: algorithms optimize for engagement, not necessarily quality. Streaming platforms often promote what keeps viewers watching, not what is most meaningful. Periodically stepping outside your algorithmic bubble and using curated human lists or award winners keeps inputs diverse.

> **Takeaway**: Train recommendation systems with intentional input, so choosing what to watch or read becomes quick and aligned with your preferences.

BRINGING IT TOGETHER

Life administration is repetitive by nature. Meals, grocery lists, birthdays, travel, documents, budgets, and leisure all go through cycles. When those tasks rely on memory or last-minute effort, they consume more attention than necessary.

The key thread spanning this chapter has been to create predictability. We discussed recurring reminders with lead time, consolidated lists, structured folders, categorized spending, and clear evaluation criteria. These adjustments are modest but cumulative, creating room for higher-value work.

That matters because the next step is output. Once administration is under control, you can direct more energy toward work, creativity, and problem-solving.

AI TOOLS REFERENCED

Before listing tools, one principle is worth repeating: tools only work when paired with a simple system. AI does not fix disorganization. It accelerates whatever structure you give it. The more consistent reminders, folders, categories, and filters, the more valuable these tools become.

- **Google Calendar / Apple Calendar**: Use recurring annual events with early planning reminders and short-term execution reminders. Built-in scheduling intelligence ensures important dates surface before they become urgent.
- **Notion AI**: Organize gift ideas, travel notes, or side hustle concepts inside structured databases. Use AI to summa-

rize entries, categorize ideas, and evaluate options against defined criteria.

- **Perplexity AI**: Conduct fast comparison research with summarized answers and cited sources. Useful for travel planning, market validation, and product evaluation.
- **Google Flights**: Track airfare trends using predictive pricing models and automated alerts. Reduces manual fare checking and improves timing decisions.
- **Airbnb / Booking.com with AI Filtering**: Use advanced filters and AI-assisted review summaries to identify consistent praise or complaints, rather than reacting to isolated reviews.
- **Google Drive / Microsoft OneDrive AI Search**: Retrieve documents based on content and context rather than file name. Embedded AI indexes document text for faster location.
- **ChatGPT / Claude**: Generate structured ideas, analyze spending patterns, model scenarios, sequence itineraries, and evaluate side hustle options using constraints and real data.
- **Monarch Money**: Automatically categorize transactions and surface spending trends. Supports structured monthly review and AI-assisted forecasting.
- **StoryGraph / Goodreads**: Train recommendation engines by consistently rating books and filtering by mood, pacing, and genre for more precise suggestions.
- **Likewise / Taste.io**: Provide structured entertainment preferences to improve cross-platform recommendations for movies, shows, and books.
- **Apple Reminders / Microsoft To Do / Google Tasks**: Set recurring and smart reminders for routine life administration tasks so they do not rely on memory.

CHAPTER 3

Work and Creativity: Getting Stuff Done

Once meals, schedules, and logistics stop demanding constant attention, the next pressure point is creation. Activities such as writing, creating presentations, drafting reports, making decisions, and communicating require thinking rather than remembering.

In this chapter, I move from life management to producing results by examining how to get started, finish faster (without lowering standards), automate what doesn't require judgment, and remove time-wasting repetition.

WAY #17: "I HAVE WRITER'S BLOCK"

Co-Writing with ChatGPT and Other Tools

You sit down to write, open a blank document, and immediately feel the weight of expectation. The idea sounded clear in your head five minutes ago. Now it feels vague, unfinished, or not quite good enough to put into words. You adjust the title, type a sentence, delete it, and then check email. Writer's block often manifests like this: hesitation and easy distraction.

Many people think writer's block is about creativity, but that's not usually the case. The brain is being asked to generate structure, clarity, language, and coherence simultaneously. That is a high cognitive demand from a cold start. When those layers stack, they compound, making it feel like you have nothing to say, even when you actually have plenty.

AI is a very powerful structural partner. You may not want to use it as a ghostwriter, but it can be a great tool for helping with structure, which can help you get over writer's block.

Start by lowering the bar. Instead of "write the article," prompt for an outline.

AI Prompt:

```
Create three different outlines for a 1,200-word article on
[topic]. One should be analytical, one narrative, and one
practical. Include section headers and 2-3 bullet subpoints
per section.
```

That step replaces a blank page with options. Once an outline is selected, expand it incrementally.

AI Prompts:

```
Expand section two into a detailed breakdown with examples.
Add one counterargument and a short response.
Suggest a practical example that illustrates this idea.
```

This prompt sequence builds layer by layer. First, generate the structure, then the substance. At no point is the model asked to "write the whole thing in my voice." You are directing the thinking.

Claude is particularly strong when working with longer drafts or being fed with existing material. If you have 1,500 rough words that feel disorganized, paste them in.

AI Prompt:

```
Reorganize this draft into a clearer structure without
removing any core ideas. Preserve tone. Identify redundancies.
```

That kind of restructuring takes significant time; it is exactly the type of compression and organization large language models handle well.

If writing with Google Docs, the built-in AI drafting tools can generate paragraph expansions inline. In Notion, a rough bullet list can be highlighted and then "expand" selected. The advantage of these tools is momentum. You don't break flow by switching contexts, but instead iterate inside the same workspace you're writing in.

Writer's block often persists because the task feels undefined. A 1,000-word article can be intimidating, but a three-part outline is not. A rough paragraph with imperfect phrasing is not. The psychological shift from "create something good" to "improve something rough" is significant. AI helps get to the "improve" stage faster, which helps speed up the writing process.

There is also a second layer to writer's block: self-criticism. Many people stop mid-draft because the writing does not sound polished yet. That is premature evaluation. Here's a useful prompt for breaking that loop.

AI Prompt:

> Generate a deliberately imperfect first draft based on this outline. Focus on clarity over style.

This can remove the expectation of polish. Editing is a different cognitive mode from drafting. Keep them separate.

For more complex writing, such as research-backed essays or business strategy pieces, use Perplexity to gather structured source material first.

AI Prompt:

> Summarize three credible perspectives on [topic] and provide links.

Then feed those summaries into ChatGPT or Claude.

AI Prompt:

```
Using these perspectives, draft a balanced analysis with
clear pros and cons.
```

This workflow separates research and synthesis into distinct steps, rather than blending them chaotically. This is generally how experienced writers think, just faster.

A deeper workflow that works consistently and helps remove ambiguity at each step looks something like this:

1. Define the outcome clearly. Who is this for? What should they understand or do?
2. Generate multiple outline options.
3. Select and refine the strongest structure.
4. Expand section by section.
5. Insert one concrete example per major section.
6. Edit for tone, clarity, and concision.
7. Cut 10 to 20 percent for sharpness.

Removing ambiguity at each step keeps things flowing, cutting down on writer's block.

If you feel stuck because your argument feels thin, adversarial prompting is a powerful technique.

AI Prompt:

> ```
> Challenge this thesis. What are the strongest counterarguments?
> ```

Then have a conversation with the AI chatbot and respond to the counterarguments. This can deepen your thinking and provide either the confidence that the argument is complete or help you fortify it. Having a devil's advocate can be extremely helpful.

I don't believe AI should replace your thinking. It should accelerate it—and your writing. If you accept output uncritically, the writing will flatten. Use AI to create the clay and then sculpt it.

Always scrutinize the output of chatbots. Large language models (LLMs) can fabricate sources, misinterpret nuance, or generalize too broadly. If writing about technical, legal, or financial subjects, verify facts independently. I recommend treating AI as a collaborator, not an authority.

Be careful not to become too over-dependent on AI tools. If you begin every piece by asking the model to decide the angle, you may weaken your framing instincts. A good habit is to write a rough thesis, even a messy one, and then use AI to pressure test or expand it.

Writer's block feels like a creativity failure; however, in most cases, it is a workflow failure. AI can help remove the blank page and keep the momentum flowing.

> **Takeaway**: Use AI to build structure and generate rough material quickly to focus energy on refining ideas instead of staring at an empty page.

My Personal Experience

At the end of 2025, I realized I had a problem. I was publishing consistently, but not always strategically. Some articles were aligned with revenue and long-term positioning, while others were simply topics I found interesting that week. Individually, they were fine. Collectively, they lacked direction.

So, I changed how I approached planning. Each quarter, I now sit down with ChatGPT and give it real context about my business. I include revenue breakdown numbers, revenue streams, site structure, which books generate the most income, which tools convert well, and what areas of the site I want to strengthen. I also explain my broader goals: more evergreen content, stronger internal linking, and clearer pathways to the books and tools that support the business.

Then I ask it to evaluate my existing editorial calendar and suggest additions or adjustments. I prompt it to identify keyword gaps, underdeveloped topic clusters, and opportunities to write articles that naturally connect to higher-value pages. I also ask for proposed outlines and suggested internal links for each recommended article.

Now, I spend less time wondering what to publish next and second-guess myself less. The calendar feels intentional. My web traffic has shifted more clearly toward revenue-generating pages, and internal linking has become easier because it's designed into the plan rather than added later. I still write articles that interest me. But now they sit inside a structure that supports the business, and I spend less time planning content.

WAY #18: "MY REPORTS TAKE FOREVER"

Generating Outlines and Summaries

Ever finished a report and felt like most of your time went into rearranging paragraphs? Reports often drag because their structure is often clarified after the writing begins. Most people create the structure as they go, inserting analysis midway, and then later discover paragraphs and sections need to be rearranged so the document makes more sense.

The fastest way to shorten the time it takes to finish a report is to separate structure from drafting. Before writing a single paragraph, define the output:

- Who is this for?
- What decision should it enable?
- What level of detail is expected?

Once that's clear, generate a framework, the "spine" of the report.

AI Prompt:

```
Create a structured outline for a quarterly operations
report to an executive team. Include metrics, analysis,
risks, and recommendations.
```

If working inside Microsoft 365, another approach is to use Copilot in Word to draft or reorganize sections directly within the document. Instead of copying content into another tool, highlight raw notes.

AI Prompt:

```
Summarize these notes into a concise analysis section
aligned with the outline.
```

If the report relies on data, Excel with Copilot can summarize trends across sheets. Rather than manually interpreting fluctuations, ask.

AI Prompt:

```
Identify significant changes compared to last quarter and
highlight anomalies.
```

For research-heavy reports, Perplexity can gather external context quickly.

AI Prompt:

```
Summarize current industry benchmarks relevant to this metric.
```

Then incorporate only what strengthens your argument.

A helpful internal workflow looks like this:

1. Draft the outline.
2. Insert raw data and notes under each heading.
3. Use AI to compress and refine each section.
4. Add one forward-looking recommendation per major section.
5. Cut redundancy.

One caution: AI-generated summaries sometimes ignore nuance. If a report includes sensitive interpretation or strategic positioning, always review for tone and implication.

> **Takeaway**: Define structure first, then use AI to compress the raw material, so report writing becomes assembly instead of reorganization.

WAY #19: "I NEED A PRESENTATION, FAST"

SLIDES AND SCRIPTING

Presentation quality suffers for many predictable reasons. Sometimes too much text, sometimes slides that try to overexplain key concepts instead of simply supporting them. When the presentation preparation time is short, those mistakes multiply.

Again, AI is a good brainstorming partner, and can help tighten up a presentation and have it laser-focused towards your goals and audience. To get started, define some context and constraints, such as how many slides to use and what the audience should walk away knowing. Once that is generated, prompt a slide architecture.

AI Prompt:

> Create a 12-slide executive presentation outline on [topic]. Each slide should have a clear headline and no more than three supporting points.

In minutes, tools like Gamma and Tome can convert outlines into visually structured drafts. They are useful for layout acceleration, especially when design is not the focus. PowerPoint's Copilot can generate slides directly from a document or prompt. After having a draft, you can further refine it.

AI Prompts:

> Reduce text density by 30 percent without losing meaning.
>
> Rewrite slide headlines to be more decisive and outcome-oriented.

Once you are happy with the slide deck, get help with the actual presentation.

AI Prompts:

> Write a concise 45-second explanation for slide five, assuming the audience is skeptical.
>
> Based on this slide outline, what objections might the audience raise?

You could use this information to build a slide or prepare talking points to address any issues the AI anticipates.

One thing to watch out for is that auto-generated decks often look polished but generic. Always adjust examples and language to reflect your context.

Takeaway: Use AI to accelerate slide structure and scripting to focus on clarity and persuasion, not formatting.

WAY #20: "MY TO-DO LIST IS NEVER-ENDING"

Automating Routine Work

Look closely at your list this week. How many items are repetitions? Sending status updates, copying information from one system to another, formatting reports, scheduling follow-ups, renaming files, rewriting similar emails... the list goes on and on. Individually, these tasks seem manageable, but collectively they can seem weighty.

High-value work requires judgment, while routine work follows rules. Rule-based tasks are the primary candidates for automation. If a task is predictable, recurring, and requires no emotional nuance, it is a "logical drain" that should be moved off your plate.

Start by auditing tasks completed over the past two weeks. Open your task manager and/or calendar history and categorize items into three buckets:

1. Judgment-heavy (requires thinking, creativity, or decision-making).
2. Repetitive but necessary.
3. Purely mechanical.

Most people are surprised by how large the second and third buckets are. Fortunately, those buckets are where automation delivers efficiency.

AI plays a role here, but it is not the only role. Many productivity gains come from combining LLMs with workflow automation tools and built-in platform features. For example, if you regularly convert meeting notes into follow-up tasks, you can:

1. Record meetings using Zoom or Teams transcription.
2. Use Microsoft Copilot or ChatGPT to extract action items.
3. Automatically send those items to Asana, ClickUp, or Todoist using Zapier or Make.

The last step can take some setup time; however, the entire chain can then run with minimal manual input. Zapier and Make are particularly useful because they connect systems that do not naturally talk to each other. A few examples:

- New client form submission → Create project folder in Google Drive → Generate onboarding checklist in Asana.
- Invoice email received → Save attachment to specific Drive folder → Notify accounting channel in Slack.
- Calendar event with "Prep Required" in title → Automatically create prep task two days before.

None of these workflows requires advanced coding. They require clarity about triggers and outcomes, along with a little experience using Zapier and Make. In fact, you can ask ChatGPT to give you step-by-step instructions on how to make a workflow.

AI Prompt:

```
Here is my weekly task list. Identify tasks that are repetitive
and suggest automation opportunities.
```

The model can often spot overlooked patterns that you have normalized.

Automation inside tools is equally powerful. In ClickUp or Asana, you can create rules such as:

- When status changes to "Approved," assign to operations.
- When due date approaches within 48 hours, notify team member.
- When form field is completed, generate subtask checklist automatically.

These small rules can relieve you of some micro-management tasks.

Another overlooked area is document generation. If you regularly draft proposals, reports, or onboarding emails, build templates first. Use Google Docs templates, Notion database templates, or Word's Quick Parts, and then layer AI refinement on top.

AI Prompt:

```
Rewrite this template to be clearer and reduce length by 15
percent.
```

The template now becomes the baseline, one that AI has improved. Now you can create documents faster because you're not starting from scratch each time.

Keep in mind that every repeated decision is a candidate for systemization. When you systemize, your to-do list shrinks automatically. You feel less of a burden and have more time to focus on other things, even though just as many to-do-list tasks are getting done as before. They are just happening faster with less attention and effort.

There is also an important distinction between automation and delegation. Automation handles rule-based execution; delegation shifts responsibility to another person. AI can assist in both, but they are different decisions.

For example, suppose you manually generate a weekly performance summary from multiple dashboards. Instead of copying numbers into a slide each week, consider:

- Connecting dashboards to a Google Sheet with live data feeds.
- Using Copilot in Excel to summarize trends.
- Auto-generating a summary paragraph using ChatGPT.
- Feeding that paragraph into a standing slide template.

What once took an hour can be reduced to ten minutes of review.

A structured automation audit might look like this:

1. Identify top five time-consuming repetitive tasks.
2. Estimate monthly time spent.

3. Determine if the task is rule-based.
4. Explore tool-native automation.
5. If none exist, connect tools via Zapier or Make.
6. Document the workflow once.

That documentation step matters. A system only saves time if you remember how it works. I've learned this the hard way. A few times, I've had to basically go through the steps to recreate automations and systems when I wanted to change them because I didn't have any documentation.

Automation requires setup time. Poorly designed automation can create a ton of errors. Always test workflows with limited data before fully trusting them. And, if working in regulated environments, confirm compliance requirements before routing sensitive information through third-party connectors. And, don't forget the documentation!

Automation should feel invisible when working correctly. If you notice it constantly, it probably needs refinement.

Takeaway: Identify rule-based tasks and automate or template them so a to-do list reflects decisions and impact, and not duplication.

My Personal Experience

In my IT Director role, my supervisor wanted me to provide weekly status reports on all the projects my department was running. I had a large department of software engineers, project managers, business analysts, service desk technicians,

> and network engineers. We had over a dozen projects going at any given time, and having to write up the status of each every week took almost a day of my time.
>
> My team provided me with weekly status reports. Each team did it differently: some sent a summary email, some used a Word form. I would copy and paste from all those sources and summarize them into an email. I soon realized that all I was really doing was recreating what was already given to me in a different format.
>
> We decided to standardize the report format and store them all in a SharePoint list with discrete fields, such as issues, upcoming and achieved milestones, risks, overall status, etc. With everyone putting information into the same location, in the same format, all it now takes is asking Copilot to create an email summary in the format my supervisor wants. I had to modify the prompt a few times to get it right, but now I review the output, make minor tweaks, and send it. Takes 30 minutes.

WAY #21: "I CAN'T THINK OF A GOOD TITLE/EMAIL SUBJECT"

IDEATION

Titles and subject lines are the "gatekeepers." They determine if a report is opened or an email ignored. Despite their importance, they are often written in the final seconds before hitting send, which leads to safe, generic phrasing that fails to capture attention.

AI will create a good subject line after you've pasted in the email's content.

AI Prompt:

> Based on this text, generate 20 subject lines under 50 characters. Keep tone professional and direct.

Maybe one will hit you as a good choice. If not, give more specific instructions.

AI Prompts:

> Make five more that emphasize urgency.
> Make five more that emphasize clarity.
> Make five more that emphasize measurable outcomes.

Tools like Jasper and Copy.ai are optimized for short-form marketing copy and A/B testing variants. If using email platforms like Mailchimp, built-in subject line assistants can predict open rate performance based on past campaigns.

For articles or reports, refine headlines with constraint prompts.

AI Prompts:

> Rewrite this title to be clearer and 20 percent shorter.
> Make this more specific without adding reducing the meaning.

Always carefully review the titles and headlines that AI generates to ensure they represent the desired tone. AI tends toward patterns that have historically performed well. That can produce safe phrasing. If distinctiveness matters, generate a few options and have AI refine them a few times to find one that works.

> **Takeaway**: Generate multiple framing options quickly to choose strategically instead of settling under pressure.

WAY #22: "MY JOB SEARCH IS EXHAUSTING"

Resumes, Cover Letters, and Interview Prep

Searching for a job is exhausting. Applying to each role involves resume adjustments, a targeted cover letter, company research, and interview prep. Doing all this can feel like a full-time job in and of itself.

AI can assist with each step of the process. Start with your resume and the job description.

AI Prompt:

```
Identify gaps between my resume and this job description.
Suggest ways to highlight relevant experience without
fabricating information.
```

You can also have ChatGPT revise your resume.

AI Prompt:

> ```
> Rewrite my professional summary to align with this role's emphasis on operational leadership and measurable outcomes.
> ```

Similarly, create a targeted cover letter.

AI Prompt:

> ```
> Draft a concise cover letter tailored to this job. Emphasize leadership, problem-solving, and measurable results.
> ```

Review every AI-generated draft to remove generic boilerplate phrasing. Recruiters can quickly identify cover letters that lack a human touch or sound overly optimized for algorithms. The goal is to use AI to find alignment between your history and the job's requirements, then inject specific, nongeneric details that only you can provide. The cover letter and resume need to represent you accurately.

Tools like Teal and Jobscan analyze a resume for keyword alignment and applicant tracking system (ATS) compatibility. They highlight where phrasing mismatches may reduce visibility. Consider using them to get your foot in the door.

If you are fortunate enough to get an interview, AI can help with that, too. Once again, give Claude or ChatGPT the job description, your cover letter, and your resume.

AI Prompt:

```
Generate 12 likely behavioral interview questions for this
role. Provide STAR-format answer outlines.
```

You can even have AI craft a good thank-you note following an interview. Try to remember a key question that came up during the interview, or a specific discussion, or one area that you would like to expound upon that you didn't in the interview, and give that to ChatGPT.

AI Prompt:

```
Integrate this key point into a thank you note that goes to
the interviewers for the [name of the job]
```

> **Takeaway**: Use AI to diagnose alignment gaps and structure customization, so repetition does not drain energy.

WAY #23: "I'M STUCK ON A PROBLEM"

BRAINSTORMING AND TROUBLESHOOTING

Some problems refuse to budge no matter how long you stare at them.

You review the constraints, rephrase the question, and try to push harder. The more effort applied, the narrower the thinking becomes. Eventually, all you're doing is defending your current approach.

Oftentimes, problems feel immovable because the assumptions on which they are built are invisible. You have already decided what is possible, what is off limits, what the goal really is, and what trade-offs are unacceptable. Those assumptions might be correct; however, when they go unquestioned, they shrink the solution space.

AI is a powerful sounding board because it does not share your professional blind spots or institutional history. It excels at "assumption auditing." While you are busy defending a current approach, the AI can simulate different perspectives or apply specific mental models to the problem.

The first step is to clearly and explicitly define the problem to ChatGPT.

AI Prompt:

```
Here is the problem. Here are the constraints. Here is what I
have tried. Here is the desired outcome.
```

Be specific. Vague prompts produce generic output; you want to be precise.

AI Prompt:

```
I lead a team of six. Two members are underperforming, but I
cannot reduce headcount. Morale is dropping. I have already
tried direct feedback and workload redistribution. Generate
five alternative strategies that do not involve termination.
```

Next, you'll want to expand on whatever the LLM outputs.

AI Prompts:

```
List unconventional approaches I might not consider.
What assumptions am I making that could be incorrect?
If I had to solve this without increasing budget, what would I try?
If I had to solve this with double the budget, what would change?
```

To get the most out of this process, go even deeper. Remember, you are using AI as a thought partner. Don't just accept its answers; interrogate them.

AI Prompts:

```
Rank these by risk.
What could go wrong with option three?
Which solution has the highest long-term upside but short-
term discomfort?
```

For more research-heavy problems, Perplexity is useful, as it pulls in external context with citations. Suppose you are evaluating whether to expand a product line.

AI Prompt:

> Summarize recent case studies of companies expanding into adjacent markets. Highlight risks and success factors.

Asking about reference material will prevent you from reinventing analysis that already exists.

In technical environments, GitHub Copilot plays a different but related role. When debugging code, the bottleneck is often pattern recognition. Copilot can suggest alternative implementations directly in your integrated development environment (IDE). Even if the suggestion is not perfect, it introduces variations you might not have tried.

So far, I've discussed various forms of brainstorming. Once again, this is also a good place to ask for a devil's advocate point of view.

AI Prompts:

> Argue against my current plan.
>
> What is the strongest criticism of this approach?
>
> If this fails, what will be the most likely cause?

Beyond simple brainstorming, use AI to perform "perspective shifting" by simulating specific stakeholders. Prompt the model to respond as a skeptical CFO focused on ROI, an operations lead worried about execution risk, or a customer who dislikes change. This allows for "red-teaming" your ideas and surfacing blind spots before a solution is presented to the team.

In general, when stuck on a problem, first clearly define it and then use AI to expand on possible solutions, evaluate trade-offs, and refine a direction. This will get you unstuck!

AI does not understand the full context unless you supply it. Additionally, organizational politics, long-term strategic vision, and interpersonal nuance are difficult to encode in a prompt. You must interpret AI's suggestions through your experience, knowledge, and expertise.

> **Takeaway**: Use AI to externalize assumptions, expand alternatives, and pressure-test solutions to move from circular thinking to deliberate action.

WAY #24: "I WASTE TIME REPEATING MYSELF"

Creating Templates and Snippets

Sometimes it feels like you're giving the same responses over and over again. The same questions. Similar onboarding instructions. Clarifying the same policy to different people.

Streamline this by identifying patterns. Review sent emails from the past month and highlight recurring explanations.

AI Prompt:

```
Extract a reusable template from these messages. Keep it concise and adaptable.
```

Refine the templates further with prompts.

AI Prompts:

```
Make this template more direct without sounding abrupt.
Shorten this by 15 percent.
```

Store the generated templates in Google Docs, Notion, or another easily accessible place.

For dynamic insertion, tools like TextExpander or Outlook Quick Parts allow the insertion of structured snippets with short triggers. Superhuman supports reusable responses for high-volume communication.

> **Takeaway**: Convert repeated communication into reusable templates to preserve time and ensure consistent communications without having to rewrite from scratch.

Bringing It Together

Work feels like it's expanding because knowledge work blends thinking, formatting, repetition, and communication into one stream. When those layers entwine, everything feels more exhausting than it needs to.

Throughout this chapter, the consistent pattern in making work and creativity easier has been to define structure first:

- Generate outlines before drafting.
- Create architecture before designing slides.
- Diagnose resume gaps before rewriting.
- Identify repetitive tasks before automating.
- Expand alternatives before choosing a solution.

Producing output is easier when the structure is defined early.

AI does not eliminate effort; it reallocates it. Instead of spending time on first drafts, formatting, and repetitive assembly, spend more time on refinement, judgment, and decision-making. You still supply the direction, which is where your value lives. Used well, AI becomes part of your production system.

AI Tools Referenced

Remember this principle: AI improves structured workflows; it does not compensate for unclear objectives or goals. The clearer your initial output, the better its ultimate output.

- **ChatGPT / Claude**: Generate outlines, restructure drafts, simulate interviews, pressure-test ideas, draft scripts, and expand or compress content with controlled prompts.
- **Microsoft Copilot (Word, Excel, PowerPoint)**: Summarize documents, analyze datasets, generate presentation drafts, and refine structure directly inside Microsoft 365 tools.
- **Perplexity AI**: Conduct fast research with summarized findings and citations. Useful for benchmarking, industry context, and external case studies.
- **Notion AI**: Expand bullet points into drafts, generate structured templates, and organize content inside databases and project workflows.
- **Gamma / Tome**: Convert structured outlines into presentation decks with automated formatting and layout.
- **Zapier / Make**: Automate cross-platform workflows such as form submissions, file routing, task creation, and notification triggers.
- **ClickUp / Asana Automation**: Build rule-based triggers within project management systems to reduce manual tracking and repetitive updates.
- **Jasper / Copy.ai**: Generate marketing headlines, subject lines, and short-form variations optimized for engagement testing.
- **Teal / Jobscan**: Analyze resume alignment with job descriptions and identify keyword gaps for Applicant Tracking Systems.
- **GitHub Copilot**: Suggest code completions and debugging alternatives inside development environments to accelerate technical problem-solving.
- **TextExpander / Outlook Quick Parts**: Store and deploy reusable templates and communication snippets to reduce repetitive drafting.

CHAPTER 4

Communication: Saying the Right Thing, Faster

Communication is where smart people often lose time. Poor communication is due to the weight and pressure that every important message carries, not by a lack of ability. You must account for tone, timing, audience, and risk. It's not just an email; it's managing perception. You must consider:

- How do I say this professionally?
- Does this sound too long?
- Can I explain this in plain English?
- Am I prepared for this conversation?

You've probably thought some version of each of these just this month.

Sometimes you may find yourself rewriting the same paragraph three times. You soften language, then strengthen it, then second-guess it again. Multiply that across emails, Slack threads, presentations, comments, and hard conversations, and communication becomes a noticeable tax on your day.

The best way to use AI is not to replace your voice, but rather to sharpen it. Used correctly, it becomes a thinking partner that helps adjust tone without losing clarity. It helps compress long messages without losing meaning, translate complexity into simplicity, and prepare for conversations that matter. It doesn't decide what you believe; it helps express that faster and more precisely.

WAY #25: "HOW DO I SAY THIS PROFESSIONALLY?"

Polishing Tone

Tone is a matter of emotional translation. You may know exactly what you want to say, but can't always predict how it will land. When frustrated, writing becomes sharp. When nervous, points are diluted with excessive qualifiers. AI provides a fast external mirror to ensure delivery matches intent. Tone is where smart people lose time.

Also, we're bad at hearing ourselves objectively. You can't fully separate the meaning in your head from the words on the page. You know your intent, so you assume it's obvious. It isn't. Worse, once something is written, you become attached to it. So, instead of refining tone strategically, you either:

- Send it as-is and hope for the best.
- Over-edit and drain the clarity out of it.
- Sit on it too long.

AI provides an instant mirror and can be a tone translator. Paste a draft with a precise instruction.

AI Prompts:

> Rewrite this email to sound professional and firm, but not confrontational. Keep it concise. Do not soften the main point.
>
> Adjust the tone to be collaborative and calm. Preserve my boundaries and deadlines.
>
> This message may sound frustrated. Rewrite it so it sounds neutral and focused on problem-solving.

The AI is not being asked to change your position, but to adjust delivery while protecting intent.

Here's a real-world example of a draft: "We've been talking about this for weeks. I still don't have the numbers. I need them by tomorrow, or this project is going to slip again."

Clear, but sharp.

An AI-adjusted version could read: "We've discussed the reporting timeline over the past few weeks, and I still haven't received the updated numbers. I need them by tomorrow to prevent additional delays on the project. Please confirm when I can expect them."

Same boundary, but different emotional temperature. That small shift happened by communicating what you need without conveying an unintended tone.

You can also ask AI to analyze tone before rewriting.

AI Prompt:

> What tone does this message currently convey? Where could it be misinterpreted?

This builds awareness. Eventually, you'll start catching patterns in your writing.

I've used this before sending sensitive messages to team members and vendors. My rule is simple: If the stakes are relational, I run it through AI once, which only takes 30 seconds and might prevent hours of cleanup.

AI doesn't know organizational politics, power dynamics, or context beyond what you paste into it. If the message is legally sensitive, HR-related, or high-stakes, use AI for tone review, not final judgment. You still own the message.

Also, don't let AI make you overly diplomatic. Professional doesn't mean vague. If the message requires clarity, protect that clarity.

AI Prompt:

> Rewrite this to sound professional, direct, and calm. Keep it under 150 words. Preserve the main request.

Compare versions; notice what changes. You'll start recognizing tone patterns faster.

Takeaway: Use AI as a tone mirror, not a ghostwriter, so relationships are protected without diluting the message.

WAY #26: "THIS EMAIL'S TOO LONG"

Summarizing and Tightening

When a message runs long, it usually means you're trying to preserve every piece of context and not leave anything out. You want to be clear, avoid follow-up questions, and so you add more explanation. Then another qualifier. Then a background paragraph that "might help."

Now the reader has to work to understand the email!

Most busy professionals ignore long emails because they're overloaded, not because they're lazy. The longer a message, the more effort it takes to extract the decision, request, or takeaway. If someone has to hunt for the main point, you've already lost them.

You reread a draft and think, "This is too long." What you're really noticing is dilution, the signal buried in noise. AI is excellent at separating them.

Here's the workflow: Write a full draft without restraint. Get everything out. Don't edit. Then paste it into AI with specific constraints.

AI Prompts:

> ```
> Tighten this email to under 150 words. Preserve all key decisions and deadlines. Remove repetition and background that isn't necessary for action.
> Summarize this into 5 bullet points focused only on what the reader needs to know or do.
> Rewrite this to be 30% shorter without losing clarity or firmness.
> ```

Notice the difference between "make this better" and "reduce by 30%." Constraints matter. AI responds well to measurable targets.

Here's a realistic example. An excerpt of an original draft: "As we discussed in last week's planning session, and again in Monday's follow-up, we're trying to finalize the Q2 rollout timeline. There are still a few open questions around vendor availability, resource allocation, and budget confirmation. I know everyone's been busy, but we need to bring this to closure so we can communicate expectations to the broader team..."

This continues for three more paragraphs.

Here's an AI-tightened version: "We need to finalize the Q2 rollout timeline this week. Open items: vendor availability, resource allocation, and budget confirmation. Please send your updates by Thursday so we can confirm the plan and communicate it to the broader team."

The shortened version doesn't lose authority, but it does gain clarity.

When I started using AI this way, I noticed something uncomfortable. My long emails weren't always about clarity. Sometimes they were about self-protection. If I explained everything, I ensured all the bases were covered. Most of the time, a shorter version was better.

You can also use AI when receiving long or multiple emails. For example, you can also use AI as a filter before meetings.

AI Prompt:

> Summarize this email thread into the three decisions we still need to make.

AI will sometimes cut too aggressively and remove nuance that matters. Always compare the shortened version to your intent. If legal, compliance, or political context is involved, protect the detail that will safeguard you. Concise does not mean incomplete.

Also, be careful not to outsource thinking. If you consistently rely on AI to figure out what matters, your summarizing skills will weaken. Use it as a compression tool after you've done the thinking, not instead of doing the thinking.

Before sending your next long message, try the following.

AI Prompt:

> Identify the single most important action in this email.
> Rewrite it so that action is clear in the first two sentences.
> Reduce total length by 25%.

> **Takeaway**: Draft freely, then use AI to compress for clarity, so the main point lands faster and cleaner.

WAY #27: "I NEED THIS IN PLAIN ENGLISH"

Explaining Complex Ideas Simply

Sometimes, even when material is understood really well, it is hard to explain to others. Being close to the subject means knowing the background, trade-offs, and edge cases. When someone says, "Can you just explain this simply?" what they're really asking is, "Can you remove everything I don't need without removing what matters?" That's harder than it sounds. The default is making sure they know everything so they understand.

Complex explanations usually fail because of the "curse of knowledge." When having a deep understanding of a subject, you might struggle to remember what it was like not to know it. You inadvertently mix definitions, edge cases, and implications into a single, confusing stream. AI helps by forcing these elements into distinct layers of information.

Here's a simple workflow I use.

AI Prompt:

```
Rewrite this explanation at three levels:

    1. Executive summary (3–4 sentences)
    2. 8th-grade reading level
    3. One-paragraph technical version
```

Then, ask for analogy support.

AI Prompt:

```
Create one clear analogy to explain this concept to someone
outside my field.
```

Analogies expose weak thinking. If the analogy fails, your understanding might not be as clear as you think.

Finally, test for hidden assumptions.

AI Prompts:

```
What background knowledge does this explanation assume?
Where might a non-expert get confused?
```

Here's a real-world example.

Let's say you're explaining multifactor authentication to a non-technical executive. The original version might sound like this: "Multifactor authentication (MFA) enhances account security by requiring additional verification factors beyond a password, such as a time-based one-time passcode or biometric validation, reducing the likelihood of credential compromise."

Accurate, but not necessarily clear in explaining how MFA improves security in practice. AI-assisted simplification could produce:

- **Executive summary**: Multifactor authentication adds a second step to logging in, such as a code sent to your phone. Even if someone steals your password, they still can't get in.
- **8th-grade version**: It's like having two locks on your door instead of one. A password is the first lock. A phone code or fingerprint is the second.
- **Technical version**: MFA combines something you know (a password), something you have (a device), or something you are (biometric data) to reduce the risk of unauthorized access.

Now you can choose based on audience.

You can also run explanations through readability tools like Hemingway Editor or Grammarly's clarity score. Hemingway will highlight long sentences and complex phrasing, which forces you to confront where the writing is doing too much at once.

AI sometimes oversimplifies. It may strip out risk conditions, tradeoffs, or important caveats. Always compare the simplified version to the original and ask: "Did anything important disappear?"

Before your next presentation or email, try this.

AI Prompt:

```
Rewrite this explanation for a non-technical audience.
Highlight any jargon and replace it with plain language.
Keep the meaning intact.
```

Takeaway: Build layered explanations with AI to adjust complexity without losing precision.

My Personal Experience

I consider it my mission to make the understanding of technical concepts easier. The motto of my business is "home technology made simple." That said, my writing is for different audiences with different experience levels. On my HomeTechHacker blog, I have articles that get deeply technical, which may not be accessible to those not in the DIY technology crowd. However, I also write for other non-technical publications where I want to cover similar topics. Instead of rewriting the articles from scratch, I often put them through ChatGPT with a description of the audience I'm targeting, and tell it to rewrite for them. It works great as a first draft, and sometimes is close to the final draft. It definitely saves me from having to write an article from scratch.

WAY #28: "I DON'T SPEAK THEIR LANGUAGE"

Quick Translations and Cultural Tone

Modern technology has made translating into another language easy. The harder part is translating intent. It's not just converting English into Spanish, French, Japanese, or Portuguese; it's trying to preserve tone, hierarchy, politeness, and clarity across different communication norms. A message that sounds confident in one culture can sound aggressive in another. A message that feels collaborative in one country can sound vague in another.

That's where AI becomes more than a dictionary.

AI Prompt:

> Translate this message into German. Maintain a professional but friendly tone.

Then go one step further:

AI Prompts:

> Adjust this for business communication norms in Germany.
> Make sure it sounds appropriately formal.
> Does this message sound too direct for Japanese business culture? Suggest adjustments.

You're asking AI to localize, not just translate. For example, consider this original English message: "I need this by Friday. Please confirm."

In some contexts, that's perfectly fine. In others, it can sound abrupt.

A localized version might become: "Could you please confirm whether you'll be able to provide this by Friday? Let me know if you foresee any issues."

The deadline remains, but the tone shifts.

AI can also be used to identify cultural assumptions embedded in your language.

AI Prompt:

> Review this message and flag any phrasing that may be culturally specific or easily misinterpreted internationally.

That prompt surfaces hidden idioms, slang, or implied expectations.

It's also useful when English isn't the other person's first language. You can simplify before translating.

AI Prompt:

> Rewrite this in plain English with short sentences and no idioms. Then translate into Spanish.

Beyond chat-based tools, platforms like DeepL often produce more natural translations. Grammarly can also flag tone issues in English drafts before translating them. If publishing or sending something public-facing, running it through both an LLM and a dedicated translation tool provides a comparison baseline.

Different cultures vary in directness, formality, and decision-making norms. Some value explicit clarity and short communication. Others expect more context and relational framing. AI can simulate these differences, but it only works as well as the instructions provided will let it.

AI Prompts:

> ```
> Rewrite this message for a high-context culture where
> indirect communication is more common.
> Rewrite this for a low-context, direct communication
> culture. Keep it concise and explicit.
> ```

However, AI does not truly understand culture. It predicts patterns based on training data. It may overgeneralize, stereotype, or exaggerate formality. If a relationship or contract is important, consider a human review, especially for legal or regulatory language.

And remember, clarity still wins. Overcompensating for cultural sensitivity can make a message so cautious that it becomes vague. Keep intent intact.

Before your next international message, try this sequence:

1. Simplify the English version.
2. Translate it.

3. Adjust tone to the target culture's business norms.
4. Flag any phrases that might be misinterpreted.

This can prevent misunderstandings that cost much more than the minute it takes to run the sequence.

> **Takeaway**: Use AI to localize tone and expectations, not just translate words, so intent survives across cultures.

WAY #29: "I WANT TO IMPROVE MY WRITING"

Personalized Feedback and Grammar Help

Most people stop improving their writing the moment school ends. After that, it becomes functional. You send emails, draft reports, and update slides. As long as it's clear enough, you move on. However, promotions, influence, and trust are built in part on written material.

Most people use AI as an advanced spellcheck. They paste a paragraph and ask it to fix grammar or make it "sound better." That's useful. It catches surface issues. But it doesn't improve you; it improves that draft.

The real leverage starts when AI becomes a pattern detector. Instead of asking, "Can you fix this sentence?" you ask, "What patterns do you see in my writing?" This will lead to better writing.

Start with three to five recent pieces of writing. Emails, report excerpts, proposal sections, even Slack messages. Paste them into AI.

AI Prompt:

> Analyze these samples and identify recurring patterns in my writing. Highlight strengths. Highlight weaknesses. Note tone tendencies, sentence length patterns, hedging language, passive voice, and clarity issues.

You may get uncomfortable insights. Maybe you over-explain, or hedge with phrases like "just," "maybe," or "I think." Maybe sentences run long and bury the main point. Maybe the tone shifts from confident to apologetic depending on the audience. In any event, you start to identify habits.

From there, go deeper.

AI Prompt:

> Based on these patterns, give me three targeted improvement exercises. Keep them practical and specific.

The exercises might include:

- Rewrite one paragraph using only sentences under 20 words.
- Remove all hedging language and compare tone.
- Move the main action request to the first two sentences.

You can also compare drafts intentionally.

AI Prompt:

> ```
> Compare Version A and Version B. Which is clearer and more
> persuasive? Explain why in concrete terms.
> ```

Notice that the last sentence of that prompt begins with: "Explain why." That's where the learning happens. AI becomes a writing coach, not just a rewriting engine.

For readability, tools like Hemingway Editor will flag dense sentences and passive constructions. Grammarly can highlight tone shifts or clarity issues. ProWritingAid can show overused words across longer documents. These tools surface mechanical patterns. An LLM helps interpret them and suggest strategic improvements.

Here's a simple monthly loop you can adopt.

- Save three recent pieces of writing.
- Run a pattern analysis prompt.
- Choose one weakness to focus on for the next month.
- Before sending important messages, run a quick review and flag any instances of a specific weakness.
- Re-run pattern analysis after 30 days.

Over time, your baseline improves.

I've noticed this in my own writing. When I started asking AI to analyze patterns instead of just rewriting drafts, it flagged rhythm issues I hadn't seen. Too many similar sentence openings. Paragraphs that ran long before the main point appeared. Once I became aware of it, I could correct it before asking for help. I'm still improving, but AI has helped.

I want to caution that AI will often smooth writing into something bland and generic. It optimizes for clarity and safety, not originality. If you accept every suggestion, your voice disappears. The goal is not to sound like a well-trained chatbot. The goal is to become more precise while still sounding like you.

Before your next important draft, try refining it first.

AI Prompts:

> ```
> Identify recurring writing patterns in this draft.
> Suggest improvements without changing my voice.
> Explain the reasoning behind each major change.
> ```

Then decide what to keep. You won't need as much correction and will anticipate it yourself.

> **Takeaway**: Use AI as a writing coach that reveals patterns so underlying skills improve, not just drafts.

My Personal Experience

Even though I've written thousands of articles, books, technical manuals, emails, reports, etc., I am still a little self-conscious about my writing. I've spent a lot of time trying to improve it, often through revisions. When I started using Grammarly, it taught me a lot about the common writing mistakes I make (e.g., missing articles and all kinds of errors with comma placement). I have learned to improve my writing based on Grammarly's suggestions.

ChatGPT allowed me to take that improvement a step further. A few years ago, I started having it analyze samples of my writing. One of the things it told me was that my paragraphs and sentences were often too long for the audiences I write for. I have tried to take that advice with first drafts. It has given me other suggestions over the years, and sometimes I think they are good, sometimes I ignore them (mostly because some suggestions subtly change meanings or take things out of my voice). So, it's an assistant in the role of an editor, and I'm free to follow the suggestions or not.

WAY #30: "I NEED TO PREP FOR A TOUGH CONVO"

Scripting with Empathy

Some conversations sit in your head for days before they happen. You rehearse them in the shower, replay possible reactions while driving, and draft sentences realizing you'll never speak them aloud. The issue is trying to balance honesty, authority, and empathy.

Difficult conversations usually have at least three layers: the desired outcome, the relationship to preserve, and emotions such as frustration, disappointment, fear, or defensiveness. Conflict occurs when these layers blur.

AI can't feel tension for you, but it can help separate those layers before walking into the room. Start by getting rid of the emotion privately by working through unfiltered thoughts.

AI Prompt:

> I need to have a difficult conversation about [topic]. Here's what I'm feeling and what I want to say. Help me clarify the core issue and the outcome I actually want.

Dump everything. Don't edit yourself. The goal is clarity, not polish.

Next, ask AI to structure it.

AI Prompt:

```
Separate this into:

    1. The core problem
    2. The impact of the problem
    3. The outcome I'm seeking
    4. The relationship considerations
```

You'll often realize you were mixing them. Now move on to the next step.

AI Prompt:

```
Draft a version of this conversation that is direct, calm, and
respectful. Preserve my boundaries. Avoid sounding accusatory.
```

Now, take it further. Ask for pushback.

AI Prompt:

```
Roleplay the other person responding defensively. Show me
how they might react and how I can respond constructively.
```

For example, imagine needing to address a team member who keeps missing deadlines.

An emotional draft might look like: "I'm tired of chasing you for updates. This keeps happening, and it's affecting the team."

AI might restructure it into something like: "I've noticed several deadlines have slipped over the past month. When that happens, it affects the team's ability to plan and creates additional work. I want to understand what's getting in the way and figure out a system that prevents this going forward."

It's the same issue, but with different framing. The focus shifts from accusation to impact to solution.

Next, simulate resistance: "I've been overloaded. No one else seems to understand that."

Now practice your response before the real conversation: "I understand workload has been heavy. If that's the case, we need visibility earlier so we can adjust resources. Let's agree on a check-in process that prevents surprises."

You can also use AI to surface empathy gaps.

AI Prompt:

> ```
> Review this script. Where might the other person feel attacked or unheard?
> ```

This approach works beyond work conversations and can be applied to family logistics, financial boundaries, and even saying no to commitments. The structure is the same: separate emotion from outcome, script with clarity, simulate reaction, and refine.

There's something deeper happening here. By rehearsing difficult conversations deliberately, you will reduce emotional volatility and be less reactive since you've already tested the terrain. AI becomes a cognitive rehearsal partner.

That said, don't over-script. Real conversations are dynamic. If lines are memorized, you'll sound mechanical. The goal is preparation, not performance. Use the script to clarify intent, then speak naturally.

Also, remember that AI predicts likely responses based on patterns. It doesn't know the full psychology of the person you're speaking to. Treat role-play as scenario planning, not prophecy.

Before your next hard conversation, try this sequence:

1. Clarify the core issue and desired outcome.
2. Draft a calm, direct script.
3. Simulate pushback and refine your responses.
4. Ask for empathy gaps in your wording.

That preparation can turn a tense interaction into a productive one.

> **Takeaway**: Use AI to separate emotion from intention and rehearse clarity so difficult conversations become deliberate, not reactive.

WAY #31: "I CAN'T KEEP UP WITH THE COMMENTS"

DRAFTING REPLIES FASTER

Communication volume has exploded. Emails, Slack threads, LinkedIn comments, blog replies, customer questions, and internal updates. None of them is individually overwhelming. Together, they fragment attention. To clear the queue, you start answering quickly, which leads to a drop in quality, including inappropriate tone slipping in.

Instead of responding one by one from scratch, batch first.

Copy five to ten similar comments or emails into AI.

AI Prompt:

> Categorize these messages by theme and identify the most common questions or concerns.

You'll usually find patterns. Clarification requests, objections, pricing questions, and technical misunderstandings. Once you see the categories, draft responses based on them.

AI Prompt:

> Draft a clear, professional response to each category. Keep it concise and adaptable. Leave room for personalization.

Now, you have templates. When a new comment comes in, paste the specific message along with your base response.

AI Prompt:

```
Personalize this response for this specific message.
Maintain warmth and clarity.
```

This works especially well for:

- Blog comments,
- Customer support replies,
- Frequently asked internal questions,
- Social media responses, and
- Proposal follow-ups.

You can also use AI to summarize long comment threads.

AI Prompt:

```
Summarize this thread into the key points raised and any
unresolved questions.
```

That alone can save the time of having to reread threads.

If publishing content regularly, build a response bank. Keep a document of high-quality answers to recurring questions. Periodically paste them into AI.

AI Prompt:

```
Tighten these responses and ensure consistent tone.
```

Communication becomes both faster and more consistent.

There is one caution here. Speed can reduce feelings of empathy and understanding. If every reply sounds identical, people notice. Always add a sentence that references something specific in the person's message. AI can help generate it, but you choose it.

Also, be careful with emotionally sensitive responses. If someone is frustrated, angry, or vulnerable, avoid template-mode. Use AI to help structure thinking, but craft the final reply with care.

Before your next comment backlog session, try this:

1. Categorize similar messages.
2. Draft base responses.
3. Personalize before sending.
4. Build a reusable response library.

> **Takeaway**: Batch, categorize, and personalize with AI so you respond consistently without burning mental energy.

BRINGING IT TOGETHER

Communication struggles often show up in subtle ways:

- You may hesitate before sending a message, over-explain, or simplify poorly.
- You may avoid a difficult conversation.
- You may let comments pile up.

Together, these issues can make communication feel overwhelming and challenging.

This chapter showed AI being used in three distinct ways:

1. As a tone mirror to help see how words might land before someone else sees them.
2. As a compression engine to separate signal from noise and make intent easier to see.
3. As a rehearsal partner and coach to help clarify thinking, improve writing patterns, and prepare for conversations that carry emotional weight.

AI does not replace judgment; rather, it surfaces options and shows patterns to accelerate drafting and restructuring. You remain responsible for clarity, empathy, and integrity.

When AI is used deliberately, communication becomes something you design. And when expressing yourself clearly and consistently, people respond differently. Although AI is powerful, its true leverage is in how you use it.

AI TOOLS REFERENCED

- **ChatGPT**: Used for tone analysis, rewriting, simplification, role-play, pattern detection, and drafting structured responses.
- **DeepL**: High-quality translation tool for natural-sounding language conversion and comparison against LLM outputs.
- **Grammarly**: Clarity, tone detection, and grammar feedback for professional writing.
- **Hemingway Editor**: Readability and sentence-density analysis tool to simplify complex writing.
- **ProWritingAid**: Pattern detection across longer documents, including overused words and structural habits.
- **Notion AI**: Summarization and rewriting within note-taking workflows.

CHAPTER 5

Family and Home: Lightening the Load at Home

Home is supposed to be where life feels lighter. Often, however, it is where small pressures accumulate. Homework questions pop up at 8:30 p.m. when everyone's already tired. Calendar conflicts turn into arguments that aren't really about the calendar. Chores drift to one person without anyone formally deciding that's how it should work. Even planning something simple for the weekend can start to feel like coordinating a small project.

Such problems are ordinary and manageable in isolation. The challenge is that they repeat, and repetition turns small stresses into steady mental weight.

AI won't raise your kids or repair a relationship, and it won't magically organize years of digital clutter with a single prompt. What it can do is reduce mental load by helping you think more clearly about routines, how to structure shared systems, generate better options, and approach recurring tensions with more intention. It delivers language when you're tired, structure when things feel messy, and perspective when you're too close to the situation.

Used thoughtfully, AI becomes a valued assistant in the background of family life. It doesn't replace conversations, but it can support them. Also, it doesn't automate relationships, but it can remove unnecessary strain.

WAY #32: "MY KID NEEDS HOMEWORK HELP"

TUTORING

Homework used to mean one of two things. Either you understood the material and could explain it, or you didn't, and everyone got frustrated. Today, there's a third option. AI can be an effective tutor, or it can become a shortcut machine. The difference depends entirely on how you use it.

The goal is to help a child understand the material in a way that builds thinking, not dependency. You don't want AI to always give them the answers.

Start by shifting the role AI plays. Don't instruct "Solve this problem."

AI Prompt:

> ```
> Explain this concept step by step as if you're tutoring a
> middle school student. Ask guiding questions instead of
> giving the answer immediately.
> ```

That small change turns AI from an answer engine into a Socratic coach.

For example, your child is stuck on a math problem.

AI Prompt:

> `Walk through how to approach this problem without solving it yet. Ask my child what the first step should be.`

Now the AI is prompting thinking. You can also adjust difficulty levels.

AI Prompts:

> `Explain this at a 5th-grade level.`
>
> `Now explain it again with more depth.`
>
> `Now create three practice problems similar to this one.`

Beyond chat-based tools, there are other useful supports. Khan Academy integrates AI-powered hints and step guidance in structured coursework. Grammarly can help older students improve writing clarity and grammar without rewriting their essays. Tools like Quizlet use spaced repetition for vocabulary and memory reinforcement.

If a child is writing, ask AI to review structure and argument strength rather than rewrite paragraphs.

AI Prompt:

> `Review this essay and identify areas where the argument is unclear. Do not rewrite it. Suggest questions the writer should consider.`

That protects ownership. Another powerful use case is study planning. Don't cram the night before a test.

AI Prompt:

> ```
> Create a 5-day study plan for this test based on these
> topics. Include practice questions and short review sessions.
> ```

You can even use it to teach metacognition.

AI Prompt:

> ```
> Ask my child three questions that help them reflect on what
> they don't understand yet.
> ```

I've seen how much calmer homework time becomes when it moves from "get it done" to "understand the process." Well-structured explanations and good instructional pacing reduce homework frustration for both the tutor and the child.

There are real concerns here, and they deserve attention. AI can make it too easy to bypass effort. If a student copies answers without thinking, then their learning decreases. Overuse can also weaken problem-solving stamina. That's why structure matters. Make it a rule that AI explains and questions before it solves. Make children attempt the problem first. Treat AI as the last step in clarification, not the first step in completion.

It's also worth checking school policies. Some teachers welcome AI-assisted learning, while others draw stricter boundaries.

Before the next homework session, try this sequence:

1. Have your child explain what they think the problem is asking.
2. Ask AI to guide, not solve.
3. Generate similar practice problems.
4. Review mistakes together and ask AI to explain the gap in reasoning.

Used this way, AI becomes a learning amplifier, not a crutch.

> **Takeaway**: Use AI as a thinking partner that guides and questions so a child can build understanding instead of outsourcing effort.

My Personal Experience

My children are older now, so we don't spend much time helping with homework anymore, but I remember plenty of the frustrations we used to have. Much of the challenge came from trying to explain concepts in ways our kids would understand. The way my wife and I learned something wasn't always the way it was being taught in school, and that gap often led to confusion for everyone involved.

Thinking about it now, AI could have helped bridge that explanation gap. Instead of trying to translate the concept ourselves, we could ask AI to explain it in simpler terms, show the steps in the way students were currently being taught, or give a few different examples until something clicked. Many

AI tutoring tools are designed to guide students through the reasoning rather than just giving the answer, which keeps the learning process intact. Used that way, AI becomes less of a shortcut and more of a patient tutor that can help both parents and kids understand the material.

WAY #33: "WE NEED A CHORE CHART"

Family Routines

Conversations about chores usually begin with someone feeling overloaded: the dishwasher didn't get emptied, the trash is still full, laundry is piling up, etc. What's framed as a small-task issue is often really about fairness and visibility. The problem is that expectations about chores are vague and memories unreliable.

A chore chart isn't about micromanaging; rather, it's about making responsibilities visible so they don't live in only one person's head. AI can help design a system faster and more thoughtfully than starting from scratch.

Begin with a brain dump.

AI Prompt:

```
List all recurring household tasks in a family with [number] adults
and [ages of kids]. Include daily, weekly, and monthly tasks.
```

You'll be surprised what shows up. Replacing air filters, scheduling dentist appointments, cleaning out the fridge. Many invisible tasks are cognitive, not physical.

Next, assign age-appropriate responsibilities.

AI Prompt:

```
Divide these tasks by age group: 6–8, 9–12, teen, adult. Suggest reasonable expectations for each.
```

Now you have a draft. From there, you can design rotation logic.

AI Prompt:

```
Create a weekly rotating chore system so responsibilities feel fair and no one is stuck with the same task every week.
```

This reduces the "Why am I always the one?" tension.

You can take things further.

AI Prompt:

```
Suggest simple rules that prevent chores from being forgotten. Keep them realistic for a busy family.
```

AI might suggest pairing chores with anchor habits, such as "trash goes out after dinner on Sundays" or "laundry starts Saturday morning before activities."

This is where tools more specialized than chatbots can help. Shared task managers like Todoist, Microsoft To Do, or Apple Reminders allow recurring task notifications. Google Sheets can track rotation schedules. Family calendar apps integrate reminders. If your family prefers visual cues, a shared Notion or Trello board can display weekly assignments. These tools all help with clarifying chores and responsibilities.

One useful approach is to have AI draft a short "family agreement."

AI Prompt:

> Write a simple, one-page family chore agreement focused on fairness, shared responsibility, and flexibility.

Something neutral can defuse defensiveness by framing chores as a shared system rather than a personal accusation.

This is also a good moment to surface invisible labor.

AI Prompt:

> Identify household tasks that are often overlooked or unequally distributed.

That prompt can lead to a productive conversation.

Resist the urge to build an overly optimized system. A chore structure that requires constant updating will collapse under real life. Keep it simple enough to survive busy weeks. Adjust quarterly, not daily.

Before the next "we need to talk about chores" moment, try this:

1. Brain dump all tasks.
2. Assign age-appropriate roles.
3. Create rotation rules.
4. Draft a short agreement that everyone can review.

You'll find that resentment decreases as expectations become clearer to everyone.

Takeaway: Use AI to make responsibilities visible and fair so that organizing chores is not just one person's responsibility.

WAY #34: "I NEED ACTIVITY IDEAS FOR THE WEEKEND"

Personalized Plans

Weekend planning sounds simple until actually trying it. You want something fun, but not exhausting; affordable, but not boring. You also want it to be close enough to be realistic, while going far enough away to feel like you did something. Add weather, age differences, and energy levels into the mix, and "What should we do?" turns into a decision fatigue loop. I've been there many times.

AI is useful here because it can filter based on constraints instead of producing a generic list. Start by giving it real parameters.

AI Prompt:

> Suggest 5 weekend activities within 30 minutes of [your city], for a family with kids ages [X and Y], budget under $50, and weather forecast of [conditions]. Include one outdoor, one educational, one free option, and one low-effort backup plan.

The specificity matters. The more constraints provided, the better the output. You can also use AI to reduce decision fatigue.

AI Prompt:

> Give me three strong options, not ten. Include one sentence explaining why each might be a good fit for this weekend.

For local accuracy, combine AI with real-world tools. Use Google Maps to verify hours and distances. Check Eventbrite or local city event calendars for current listings. Many cities publish weekend-event newsletters. AI can synthesize these.

AI Prompt:

> Summarize these local event listings and highlight the best family-friendly options.

Weather apps can also be incorporated.

AI Prompt:

> Based on this weekend's forecast, suggest indoor or covered alternatives if rain cancels outdoor plans.

If you want to stretch this further, ask AI to design a themed weekend.

AI Prompts:

> Create a "low-cost adventure day" theme with morning, afternoon, and evening activities that feel connected.
>
> Design a tech-free Saturday plan that still feels engaging.

Another powerful angle is ongoing personalization. If you consistently ask AI to generate ideas based on your family's preferences, a running "activity profile" can be built. Keep a short list of what worked and what didn't.

AI Prompt:

> ```
> Based on these past activities we enjoyed, suggest three new ones we haven't tried.
> ```

Eventually, the suggestions will improve.

Of course, no AI truly knows a local community in real time. It may suggest businesses that are closed or events that aren't happening anymore. Always verify details before heading out. Treat it as an idea generator, not a booking engine.

Also, be careful not to outsource all spontaneity. If every weekend requires a structured plan, you may create more pressure. Some of the best days are unplanned. AI is better when you're stuck, not when you're relaxed.

Next time you hear, "What are we doing this weekend?" try this:

1. Define constraints: budget, time, weather, energy level.
2. Ask for three curated options.
3. Pick one quickly and move on.

Sometimes, the biggest win isn't the activity; it's ending the decision loop and resulting fatigue.

> **Takeaway**: Use AI to filter and personalize options so planning takes minutes instead of draining half your Saturday.

WAY #35: "I WANT TO HELP MY PARENTS WITH TECH"

Translating and Teaching

Helping parents with technology can feel like stepping into two roles at once. You're the child, but suddenly also the authority figure. The questions are often small:

- "How do I update a phone?"
- "Why won't the printer connect?"
- "Is this text message a scam?"

But the emotional impact can be significant. No one enjoys feeling confused or left behind.

The issue is often a translation gap. You understand the systems; they understand their habits. The disconnect happens in language. Terms like "browser cache," "app permissions," or "Wi-Fi network" may seem obvious to you, but abstract to them. AI can act as a neutral translator between generations.

Start by simplifying explanations.

AI Prompt:

```
Rewrite these instructions for someone in their 70s who is not
tech-savvy. Use short sentences. Avoid jargon. Include step-
by-step guidance.
```

Walk them through something specific.

AI Prompt:

> Create a numbered checklist for updating an iPhone, written in plain language with no assumptions about prior knowledge.

Ask AI to anticipate confusion.

AI Prompt:

> What parts of these instructions might be unclear or intimidating for someone unfamiliar with smartphones?

This prompt forces you to slow down and remove hidden assumptions.

For more complex tasks, consider creating a printable guide. Paste the final instructions into Google Docs and format with large text and screenshots. Tools like Canva can help design simple visual guides. If the task involves identifying scams, ask AI to generate a one-page “How to Spot Suspicious Messages” cheat sheet tailored to their common devices.

Another helpful approach is rehearsal.

AI Prompt:

```
Explain this concept in three different ways:

    1.  Using an analogy
    2.  Using step-by-step logic
    3.  Using a simple real-life example
```

Now you're prepared to pivot depending on how they respond.

When instructions are calm, structured, and patient, the interaction feels collaborative instead of corrective. This is true in most learning situations, but especially important given the family dynamics that are often involved when helping parents.

There's also a cybersecurity angle.

AI Prompt:

```
Analyze this message and identify signs it may be a scam.
Explain your reasoning in plain language.
```

That reasoning step is important. It teaches pattern recognition rather than blind trust.

At the same time, be mindful of overreliance. You don't want to become the permanent intermediary between parents and every device they own. Use AI to create repeatable guides they can refer to. Encourage them to try first before calling. Consider taking the time to teach them how to use AI chatbots like ChatGPT to help themselves.

Technology changes constantly, and no single explanation will cover everything. Keep guides updated periodically, and verify instructions against the latest device software. Official support pages from Apple, Microsoft, or Google are useful cross-checks before sharing guidance.

Before your next tech help session, try this:

1. Draft simplified instructions.
2. Identify likely confusion points.
3. Create a printable or shareable guide.
4. Teach the reasoning, not just the steps.

Although the primary goal is to fix the problems, ultimately, you also want to reduce anxiety about the next one.

Takeaway: Use AI to translate complex tech into clear, respectful guidance so help feels empowering instead of overwhelming.

WAY #36: "WE'RE FIGHTING ABOUT SCHEDULING AGAIN"

Mediating with a Neutral Calendar

Scheduling arguments are rarely about the items in the calendar themselves:

- One person thought an event was optional.
- Another thought it was locked in.
- Someone else assumed childcare was handled.
- Someone else assumed that the weekend was free.

When schedules aren't clear, discussions sometimes turn into fairness arguments.

The pattern usually is repeated confusion, small resentments, reactive fixes, and no shared system that prevents it from happening again the next time. What's needed is not better memory, but a neutral calendar that is independent of and doesn't favor any people affected by it.

AI can help extract patterns from conflict before you design the solution. Start with a private reflection exercise. Paste a few recent scheduling disagreements into AI.

AI Prompt:

> Analyze these situations and identify recurring themes. Are conflicts about communication timing, unclear priorities, last-minute changes, or workload imbalance?

You're looking for root patterns, not isolated incidents.

Next, clarify shared priorities.

AI Prompt:

> ```
> Help me define 5 scheduling principles for a two-adult household
> with children, where both adults work and weekends are limited.
> ```

The response may be principles like:

- Major commitments must be entered into the shared calendar within 24 hours.
- No recurring weekend commitments without explicit agreement.
- Personal time is scheduled, not assumed.
- One person is not responsible for remembering everything.

Here's where non-chatbot tools matter. A shared digital calendar is foundational. Google Calendar, Apple Calendar, or Outlook can all work; the key is how you use them.

The first step is setting them up:

- Separate color-coded calendars for each adult.
- A shared "Family" calendar for joint commitments.
- A shared "Logistics" calendar for recurring tasks like school pickups or practices.
- A visible rule that events without a calendar entry do not exist.

If one partner prefers task-based systems, integrate Todoist, Microsoft To Do, or Things 3 with calendar views. Many of these tools allow recurring tasks and shared project boards.

After establishing a calendar, add AI into the system layer as the mediator. Export the next two weeks of calendar events and paste a summary into AI.

AI Prompt:

```
Review this schedule and identify potential overload days or
imbalance between partners.
```

AI can flag clustering you may not see.

You can also draft a "calendar agreement."

AI Prompt:

```
Write a short, neutral household scheduling agreement based on
these principles.
```

Read it together and adjust it. Treat it like a living document.

Recurring tension around personal time can be tackled.

AI Prompt:

> ```
> Suggest a fair rotation system for protected personal time in a dual-working-parent household.
> ```

If scheduling conversations often escalate emotionally, rehearse the reset.

AI Prompt:

> ```
> Draft a calm conversation starter about improving our shared calendar system without assigning blame.
> ```

You can also use project management tools if the household has complex logistics. A shared Trello or Notion board can display upcoming commitments in visual form. Some families prefer weekly planning meetings. AI can help structure those.

AI Prompt:

> ```
> Create a 20-minute weekly family scheduling agenda that covers upcoming events, workload balance, and potential conflicts.
> ```

Here's something subtle but powerful: Ask AI to explicitly play mediator.

AI Prompt:

> ```
> Based on these two perspectives, summarize each person's
> likely concerns in neutral language.
> ```

That said, no calendar fixes deeper relational patterns. If conflict stems from unspoken expectations about fairness, respect, or workload distribution, technology won't solve that; it will only expose it. The system works best when people agree to follow it. Without shared buy-in, even the most optimized calendar becomes another source of tension.

Before your next scheduling disagreement, try this sequence:

1. Identify recurring patterns.
2. Define shared scheduling principles.
3. Implement a visible shared calendar structure.
4. Review upcoming weeks for imbalance.
5. Hold a short weekly reset conversation.

> **Takeaway**: Use AI and shared calendar systems to shift scheduling from personal memory battles to an agreed-upon structure.

My Personal Experience

My wife and I have been fortunate not to run into a lot of conflicts with scheduling. We already use a shared Google Calendar for family events like doctor's appointments, kids' activities, get-togethers, and anything else that might involve more than our personal schedule. The challenge is that our kids rarely check the calendar. They'll ask if they can do something, or forget about an activity entirely, without ever looking at the schedule that's supposed to keep everyone aligned.

The AI opportunity isn't to manage the calendar but to reinforce the system around it. I could use it to generate a short weekly summary of upcoming events that goes out to the family group chat so everyone sees the same picture of the week. It could also review the schedule and flag days that look overloaded or likely to create transportation conflicts. Another useful idea is to ask AI to help draft a simple set of family scheduling guidelines so expectations about checking the calendar are clear. These are small adjustments, but they can make a calendar system work as it was intended.

WAY #37: "I NEED TO ORGANIZE OUR FILES"

Shared Systems within an AI Structure

Most families don't decide to create a messy digital system. It usually happens gradually with documents stored on desktops, unstructured folders, emails, and texts. A shared drive may technically exist, but no one consistently uses it in the same way. When needing something urgently, you search three devices and five folders before giving up.

The result is that you don't know where things are, whether they're current, or if they're backed up. Organizing family files is less about tidiness and more about reliability. Before creating folders, you need a map.

Start by inventorying what you have.

AI Prompt:

> Create a categorized list of common digital files in a household with children. Include financial, medical, school, home, legal, and personal categories.

Now compare that to reality. Where do those files currently reside? Laptop? Phone? Email inbox? Cloud storage? External hard drive?

Next, design a structure before touching anything.

AI Prompt:

> Based on these categories, design a simple shared folder hierarchy for Google Drive (or Dropbox / OneDrive) that is easy for non-technical family members to understand.

AI will usually generate something like:

- 01_Financial
  - Taxes
  - Insurance
  - Banking
- 02_Medical
  - Insurance Cards
  - Prescriptions
  - Appointment Notes
- 03_School
  - Year
  - Forms
- 04_Home
  - Mortgage / Lease
  - Utilities
  - Repairs
- 05_Legal
  - Wills
  - IDs
- 06_Photos & Memories

The key is clarity, not cleverness. Use plain language.

Now go deeper.

AI Prompt:

> ```
> Suggest file naming conventions for household documents so files
> are easy to search and sort.
> ```

A simple rule like:

- YYYY-MM-DD—Document Type—Name

makes finding the right feel much easier. For example:

- 2026-03-10—Insurance Policy—Home Renewal.pdf

NON-CHATBOT AI TOOLS CAN BE POWERFUL

Google Drive AI Search and "Ask Drive"

Google Drive includes AI-powered searching that can carry out tasks like:

- Find the last dentist invoice.
- Show documents related to 2024 taxes.

Once the folder structure and naming conventions are consistent, these AI search tools become even more accurate.

Microsoft Copilot in OneDrive/SharePoint

If your household uses Microsoft 365, Copilot can summarize document contents, identify duplicates, and do things like:

- List the insurance policies we currently have.
- Summarize our most recent mortgage document.

That only works well if documents are centralized and clearly labeled.

Dropbox Dash (AI-powered search)

Dropbox Dash uses AI to search across tools, not just Dropbox. If your family stores files in multiple places, it can unify discovery.

AI-powered Duplicate Detection

Tools like Gemini within Google Photos or Apple Photos use AI to detect duplicate photos, blurry images, and similar shots. Instead of manually reviewing 4,000 pictures, you can:

- Sort by "duplicates,"
- Remove near-identical photos, and
- Create auto-curated albums.

Now add a second AI layer: document summarization. Upload large PDFs like insurance policies or contracts.

AI Prompt:

> ```
> Summarize key terms, renewal dates, and obligations in this document.
> ```

Then, create a summary file saved alongside the original. That way, the next time you need a detail, you're not re-reading 27 pages.

You can even generate a "Household File Guide."

AI Prompt:

> ```
> Create a one-page guide explaining where major document categories are stored and how we name files.
> ```

Share it with your partner. Save it in the root folder.

Keep in mind that backups are critical. AI won't protect from hardware failure. Use any of the following:

- iCloud backup for Apple devices.
- Google Drive sync.
- OneDrive backup.
- An external encrypted drive for annual archive copies.

Then verify the backups.

AI Prompt:

> ```
> Create a quarterly checklist to ensure backups are working
> and storage limits aren't exceeded.
> ```

Automation can help here. Google Drive and OneDrive allow automatic photo and document syncing. Set it once so you don't have to rely on memory.

You may be tempted to over-engineer this. Complex nested folders and rigid taxonomies feel satisfying in the moment, but collapse under the realities of life. A system that requires perfection will fail during busy seasons. Keep it simple enough that anyone in the household can use it without asking.

Another risk is security. Centralizing documents increases convenience but also sensitivity. Enable such strategies as:

- Two-factor authentication on cloud accounts.
- Shared access permissions are limited to trusted users.
- Encrypted storage for especially sensitive documents like passports or Social Security cards.
- Occasionally, audit who has access to what.

Before reorganizing everything:

1. Inventory your categories.
2. Design a simple folder hierarchy.
3. Establish naming conventions.
4. Centralize and back up.
5. Create a one-page usage guide.

Takeaway: Use AI to design simple, searchable structures so that household files become a reliable infrastructure rather than a pile of digital clutter.

My Personal Experience

Our family stores all files in a network-attached storage (NAS) device. We have a folder structure that is descriptive (finances, personal folders for everyone, photos, videos, home movies, etc.) and flexible. However, we have a ton of files built up over decades, and finding things can be difficult, even with a known folder structure.

One way AI could improve this system is by making the files easier to find rather than reorganizing everything. Newer NAS platforms and search tools include AI features that can index files and allow for natural-language searches, such as "Tax documents from 2019" or "The video from a school play." If we ever upgrade our NAS, choosing one with AI-powered search and tagging could make decades of files much easier to navigate. Instead of relying entirely on folder memory, the system could help surface what we're looking for based on context and content.

BRINGING IT TOGETHER

Home life rarely falls apart in dramatic ways. It drifts as responsibilities blur, schedules overlap, and files scatter. Sometimes, homework sessions turn tense, with weekend planning leading to frustration and decision fatigue. None of these issues is catastrophic, but they add up to disharmony.

The goal is relief from conflict more than efficiency.

I've covered using AI to:

- Act as a learning guide so homework time builds understanding instead of frustration.
- Make chores more visible so expectations are not stored in one person's head.
- Filter weekend options instead of drowning in them.
- For aging parents, translate technology into respectful language.
- Move scheduling tension into a shared structure.
- Turn digital clutter into reliable systems.

What connects all of these is clarity around responsibilities, expectations, where things live, and how decisions are made. When systems exist, our memories can stop carrying the entire burden. AI doesn't fix relationships, but it supports the structures around them.

AI Tools Referenced

- **ChatGPT (or similar LLMs)**: Used for tutoring guidance, chore system design, scheduling analysis, document summarization, conflict pattern identification, and structured household planning.

- **Khan Academy**: Structured learning platform with AI-assisted hints and practice exercises to support homework without giving direct answers.
- **Grammarly**: Writing clarity and structure feedback for student essays without rewriting ownership away from the student.
- **Quizlet**: AI-assisted flashcards and spaced repetition for vocabulary and memory reinforcement.
- **Google Calendar / Apple Calendar / Outlook**: Shared scheduling platforms for visible commitments and color-coded household structure.
- **Todoist / Microsoft To Do / Things 3**: Task management systems with recurring task logic for chore rotation and logistics.
- **Trello / Notion**: Visual household boards for shared planning, scheduling agendas, and recurring systems.
- **Google Drive / Dropbox / OneDrive**: Centralized cloud storage systems for structured digital file organization.
- **Google Drive AI Search / Microsoft Copilot / Dropbox Dash**: AI-powered document search and summarization tools that improve retrieval and reduce digital uncertainty.
- **Google Photos / Apple Photos (AI sorting features)**: AI-assisted duplicate detection, categorization, and memory organization.
- **Canva / Google Docs**: Simple design and formatting tools for printable tech guides and shared household documentation.

CHAPTER 6

Growth and Goals: Using AI to Level Up

Most people want to improve something about their lives. Maybe it's building a new habit, learning a skill, or making progress on an idea that's been lingering for years. However, it is challenging to find enough clarity, consistency, and focus to make progress real.

Modern life doesn't always make it easy. There are more ideas, resources, and competing priorities than ever before: books pile up on nightstands, podcasts go unheard, and productivity apps remain unopened. With so many inputs, it's easy to feel like you're working on self-improvement while remaining in the same place.

AI can change that dynamic. Instead of being just another source of information, it can act as a thinking partner, coach, and sometimes even a gentle form of accountability. It can help clarify goals, break these into manageable steps, reflect on progress, and filter the constant flow of ideas into something more focused. Used like this, AI doesn't replace discipline or effort; rather, it supports the systems that make growth sustainable.

AI can:

- Help you build habits,
- Stay motivated when progress slows,
- Earn more efficiently, and
- Assist with focus.

It can be your partner in designing systems that move your life forward with intention.

WAY #38: "I WANT TO BUILD A HABIT"

Coaching and Accountability

It's not a struggle knowing which habits would improve our lives. The challenge is finding a way to build them consistently. Reading a little each day, practicing a skill, learning a language, or studying a topic all sound simple in theory. In practice, these activities disappear when life gets busy or energy runs low.

AI can help; not by replacing discipline, but by acting like a coach that supports good habits. Instead of constantly relying on motivation, AI can design a habit system, then check on progress, and adjust routines when they slip.

Start with habit design. Many habits fail because they are too vague. "Learn more about investing" or "read more books" sound great, but they don't tell you exactly what to do today. AI can help turn habits into clear routines.

AI Prompt:

> Help me design a daily learning habit that takes about 15 minutes and focuses on understanding personal finance. Suggest a simple daily structure I can follow.

You might receive something like:

- Day 1: Read a short article or book section.
- Day 2: Summarize the key idea in your own words.
- Day 3: Apply the idea to a real-life example.
- Day 4: Review and reflect on what you learned.

This won't create a perfect system, but will establish a routine small enough to sustain.

AI can also help remove decision fatigue. Instead of wondering what to study every day, ask AI to generate a plan for the day.

AI Prompt:

> Give me a 10-minute learning prompt about investing that builds on what I learned yesterday.

Another helpful use is daily accountability. Open a chat each evening and record what you did; the AI can act as a reflection partner.

AI Prompt:

> `I spent 10 minutes reading about index funds today. Ask me two questions that help reinforce the key ideas.`

Such short reflections deepen retention and reinforce the habit.

You can also build simple habit review loops. Once a week, paste notes or reflections into your AI.

AI Prompt:

> `Review this week's learning notes and summarize the three most important ideas I absorbed.`

While chatbots are the most obvious tool here, several other AI-powered tools can also support learning habits. Notion AI or Mem AI can summarize notes and highlight key ideas. Tools like Readwise can resurface highlights from books and articles so that important ideas do not disappear. Transcript tools for podcasts or YouTube videos can convert passive listening into structured learning material.

These tools work best when supporting a habit that already exists. They add reinforcement, but they cannot substitute for showing up. If the habit becomes too complex, with too many tools or steps, it will likely collapse under its own weight. The most sustainable habits tend to be the simplest.

To try this approach, start small. Choose a topic. Generate the first prompt by asking AI to help design a 10 to 15-minute daily learning routine, then commit to doing it for a week. At the end of the first week, ask AI to summarize what you learned and suggest the next steps.

Remember, progress usually comes from small actions repeated consistently.

> **Takeaway**: Use AI as a simple coaching and reflection partner to turn learning goals into small daily habits that are easier to sustain.

WAY #39: "I NEED HELP STAYING MOTIVATED"

DAILY CHECK-INS OR JOURNALING

Sometimes, you just don't wake up with the energy to do what you planned to do. Or there's too much monotony, or just dread about what's planned. Maybe it's hard to be motivated because the goals seem distant.

Most people assume the solution is more discipline or inspiration. In practice, motivation often improves when simply reconnecting with progress and purpose. Reflection helps you notice small wins, understand setbacks, and reset direction before discouragement builds. AI can support this process by acting as a simple thought partner that helps you reflect regularly.

One effective approach is a short daily check-in. Instead of letting the day pass without thinking about your purpose that day, spend a few minutes answering structured questions. AI can generate prompts and adapt them moving forward.

AI Prompt:

```
Ask me three short questions that help me reflect on progress
toward my goals today.
```

You might receive questions such as:

- What is one thing you did today that moved you forward?
- What challenge slowed you down today?
- What is one small step you could take tomorrow?

These take only a few minutes to answer, but help convert vague feelings into clearer awareness. They also reinforce a sense of momentum.

AI can also help when motivation dips. Instead of stewing in frustration, ask it to help reframe the situation.

AI Prompt:

```
I feel stuck with my goal this week. Help me analyze what might
be causing the slowdown and suggest one realistic next step.
```

Journaling is another powerful way to maintain motivation. Many people want to do so but struggle with the blank page. AI can remove that barrier by generating prompts that guide reflection.

AI Prompt:

```
Give me a short journaling prompt about personal growth that
I can answer in five minutes.
```

Prompts might include:

- What goal matters most to you right now, and why?
- What small win from the past week deserves recognition?
- What would meaningful progress look like over the next month?

These entries create a record of thinking and growth. Reviewing past entries can be surprisingly motivating, as they show how perspective and progress have evolved.

In addition to chatbots, several AI-supported tools can make this process easier. Digital journaling apps such as Day One or Journey allow you to record daily reflections and set reminders to create consistent habits. Some versions now include AI features that can summarize past entries or highlight recurring themes. After writing entries for a few weeks, ask the tool to generate a summary of what it notices about your focus, challenges, or progress.

Note-taking platforms like Notion AI can also support reflective habits. You can keep a daily journal page in Notion and periodically ask the AI assistant to analyze the entries.

AI Prompt:

> Review my last 30 days of journal entries and identify patterns in my motivation, productivity, and energy.

This type of analysis can reveal insights hard to notice day to day.

If writing feels too slow or tedious, voice tools can also help. Apps that combine voice dictation with AI transcription allow you to speak short reflections into your phone during a walk or commute. The transcript becomes a journal entry that can be reviewed or summarized later. Many chatbot platforms, such as ChatGPT, also allow voice prompts to be dictated. You can also use voice to journal and reflect with these tools.

As with habits, simplicity matters. A reflection practice that takes three minutes a day is far more sustainable than a complex system that requires 20. The goal is to maintain a regular conversation with yourself about where you are going and how you are doing. This will help you stay motivated.

Try this today:

- Ask AI to generate three daily reflection questions about your goals.
- Each evening for one week, spend three minutes answering them.
- At the end of the week, paste your answers into AI and ask it to summarize patterns in your progress and challenges.

> **Takeaway**: Use AI as a reflection partner to help notice progress, process setbacks, and maintain momentum toward goals.

WAY #40: "I WANT TO LEARN FASTER"

Summarizing Books, Podcasts, and Lessons

Books, articles, podcasts, YouTube videos, and other online courses provide plenty of learning opportunities. The challenge is turning all that information into knowledge that you actually remember and use.

Most people consume far more than they retain. You might read a great book or hear a powerful idea on a podcast, only to realize a few weeks later that you remember very little of it. AI can help close that gap by processing, summarizing, and revisiting what you learn.

A simple approach is using AI to summarize material you are already consuming. For example, after finishing a podcast episode or reading a book chapter, paste notes or a transcript into AI.

AI Prompt:

```
Summarize the three most important ideas from this material
and explain them in simple terms.
```

This summary becomes a quick reference that reinforces the key points. You can go a step further by asking AI to extract insights that are personally relevant.

AI Prompt:

> `Based on this summary, what are two ideas I could apply in my life?`

This turns passive learning into something more active.

AI can also help question what you are learning, which improves retention. Instead of just reading a summary, AI can formulate questions.

AI Prompt:

> `Ask me three questions that test my understanding of these ideas.`

Answering these forces you to think through the material. That kind of retrieval practice is one of the most effective ways to strengthen memory.

Several AI-powered tools can help turn information into usable knowledge. For example, Readwise and Readwise Reader automatically collect highlights from books, articles, and Kindle reading. AI features can then summarize these highlights or surface the most important passages. Instead of disappearing, notes become a growing knowledge base.

Note-taking platforms like Notion AI allow you to store learning notes and generate summaries or action steps directly. After finishing a book, you might paste highlights into a Notion page.

AI Prompt:

> Summarize the core ideas from these notes and suggest three practical takeaways.

Podcast and video transcripts can also be powerful learning tools. YouTube and many podcast apps provide transcripts, which can be pasted into AI tools to generate summaries, outlines, or study questions. This makes it easier to revisit complex discussions without listening to the entire episode again.

To go further, create a simple learning pipeline. Each time you read something interesting or hear an insightful podcast segment, capture the key ideas in a note. Once a week, review the notes with AI, asking it to identify themes or patterns in what you have been learning. AI summaries are most useful when they support your reflections. Reading, summarizing, questioning, and applying ideas all work together to turn information into understanding.

Without some way to process what you consume, most information fades quickly. Short reflections or summaries start to accumulate into knowledge that can actually be used. The next time you finish a podcast, article, or book chapter that feels insightful, take a minute to capture the key ideas and ask AI to summarize the three most important takeaways and generate a couple of questions that test your understanding. That small step turns passive consumption into active learning, which helps useful ideas stick.

> **Takeaway**: Use AI to summarize, question, and revisit information that has been learned so it turns into knowledge that can be remembered and applied.

WAY #41: "I HAVE TOO MANY IDEAS AND NO FOCUS"

Clarifying Thought

Sometimes the problem isn't inspiration, it's too much inspiration. The ideas just keep coming: new projects, skills to learn, systems to build, books to read, and directions to explore. The excitement of possibility is real, but so is the frustration that comes with trying to pursue too many things at once. I personally run into this problem often.

When everything feels interesting, it becomes difficult to decide what deserves immediate attention. You might start several projects and finish very few of them. Or keep collecting ideas without ever committing to the next step. Eventually, that kind of overload can make progress feel scattered—even when you are working hard.

Instead of trying to sort through all the ideas in your head, use AI to move them into a conversation, where they become easier to organize, evaluate, and prioritize.

A useful starting point is a simple brain dump. Write down every idea that has been sitting in your mind or notes: projects to start, skills to develop, topics to explore, improvements to make in your life. Once you have the list, ask AI to help organize it.

AI Prompt:

> Here is a list of ideas I have been thinking about. Group them into themes and identify which ones seem related.

This kind of prompt helps surface patterns that are easy to miss when ideas are scattered across notebooks or apps. You might discover that several ideas fall under the same theme, such as improving your health, building a creative project, or learning a new area of knowledge.

The next step is narrowing the field. Once ideas are grouped, AI can help you evaluate them more clearly.

AI Prompts:

> Based on these ideas, which three look most realistic to pursue in the next three months?
> Help me rank these ideas based on impact, effort, and how well they align with my current goals.

This does not mean letting AI decide direction. It simply provides another perspective that can help you think through options more deliberately.

AI can also help turn a broad idea into a practical starting point. Projects often stall because the first step feels unclear. Choose an idea to pursue.

AI Prompt:

> Help me break this idea into the first five steps I should take to get started.

Some tools support this kind of idea management. Notion AI allows you to capture ideas quickly and then ask the AI assistant to summarize, categorize, or analyze them. Tools like Mem AI automatically connect related notes, helping you see relationships between ideas. If you prefer capturing ideas verbally during a walk or commute, voice transcription apps can be useful. Spoken notes can later be summarized and organized by AI.

The key advantage of these tools is that they create an external thinking space. Instead of relying on memory or scattered notes, ideas become something you can examine and shape. Just as important, clarity often comes from deciding what not to pursue.

AI Prompt:

> Looking at this list, which ideas might be better saved for later so I can focus on one or two right now?

The next time you feel overwhelmed by ideas, try turning that energy into a short, structured exercise:

1. Capture every idea currently on your mind.
2. Ask AI to group them into themes.
3. Identify the two or three ideas that matter most right now.
4. Ask AI to help outline the first few steps for the one you choose.

Ideas are valuable, but progress comes from choosing one and moving forward. AI can help slow down the flood of possibilities long enough to make an intentional choice.

> **Takeaway**: Use AI as a thinking partner to organize ideas, identify what matters most, and turn possibilities into clear next steps.

My Personal Experience

Fortunately, and unfortunately, I tend to generate more ideas than I have time to pursue, especially in areas I care deeply about. My HomeTechHacker business is a good example. It allows me to pursue my passion for helping people make the most of their technology, a passion that constantly produces new ideas.

At times, it has been difficult to decide where to focus. Should I write an article, review a product, explore a partnership, or add a new feature to the site? The list of possibilities keeps growing, but my available time does not. Like many creators, I've learned that having ideas isn't the hard part; choosing which ones deserve immediate attention is the real challenge.

To bring some structure to that process, I started using ChatGPT as a thinking partner. I gave it background information about my business, including finances, goals, existing articles, and current features, along with an extended list of ideas I had been considering. I asked it to help prioritize them based on potential impact, time and resources required, and possible revenue.

I didn't blindly follow every recommendation, but the exercise helped me step back and look at ideas more objectively. It gave me a clearer picture of which projects were most aligned with my goals and which were better saved for later. That simple process helped me become more deliberate about where I invested my time and energy.

If regularly juggling more ideas than you have time to pursue, AI can provide that same kind of structured thinking space.

WAY #42: "I WANT TO AUTOMATE MY LIFE"

Building a Custom AI Assistant

Gradually building small systems that make daily life easier leads to better outcomes than reacting to whatever appears next. Developing routines for planning your week, ways to capture ideas, and structures to reduce decision-making leads to more efficiency and productivity.

Systems don't appear overnight. They emerge slowly, as people notice recurring problems and decide to solve them once, instead of repeatedly. AI makes this kind of system-building easier than ever.

Most people use AI as a tool to answer questions or generate content. That is useful, but it only scratches the surface. AI becomes far more powerful when treated as a custom assistant designed for a specific role in your life.

Instead of one general-purpose chatbot, create multiple assistants that each help with a different area of life. These assistants become small systems that support how you think, plan, learn, and make decisions.

Some common examples include:

- An executive coach that helps think through decisions.
- A fitness coach that reviews workouts and suggests adjustments.
- A learning assistant that summarizes books and podcasts.
- A planning assistant that helps organize your week.
- A research assistant that compares options before major purchases.
- A journaling partner that helps you reflect on goals and progress.

Each assistant is essentially the same AI technology, but configured with different instructions, context, and expectations.

EXAMPLE: A PERSONAL EXECUTIVE COACH

Many professionals benefit from having a coach or mentor to help them think through decisions. AI can act as a lightweight version of that support.

AI Prompt:

```
You are my executive thinking partner. Your role is to help
me analyze decisions, challenge assumptions, and identify the
most strategic next steps.
```

You can then add context about yourself:

- Your career goals,
- Your strengths and weaknesses, and
- The types of decisions you frequently face.

AI Prompt:

> `I work in technology leadership. I want to improve my strategic thinking, communication, and decision-making. When I present a problem, ask clarifying questions before suggesting solutions.`

With that set up, you can bring real situations into the conversation.

AI Prompt:

> `I'm deciding whether to pursue this new project. Help me evaluate the pros, risks, and long-term impact.`

Moving forward, the assistant becomes more useful as it learns how you think and what kinds of decisions you care about.

You might find it helpful to use the voice features of chatbots to just talk and have the text captured. This can make the coach feel more real as you talk to it.

Example: A Fitness Coach

Another useful assistant might focus on health and fitness. Many people already track workouts, nutrition, and sleep, but the data often sits unused.

You could design a fitness assistant.

AI Prompt:

> You are my fitness coach. Help me analyze my workouts, track progress, and suggest improvements based on consistency and recovery.

Then periodically share your activity data or workout summaries.

AI Prompt:

> Here are my workouts from the past week. Identify patterns and suggest one improvement I could make next week.

The assistant might highlight things such as:

- Imbalances between strength and cardio,
- Lack of recovery days, and
- Gradual performance improvements.

Some tools make this easier. Fitness platforms and wearable devices often allow data exports that can be pasted into AI for analysis. Even simple notes about workouts can provide enough information for useful feedback.

Example: A Learning Assistant

If you regularly read books, listen to podcasts, or take courses, create a dedicated learning assistant.

AI Prompt:

> You are my learning assistant. When I share notes from a book, podcast, or article, summarize the key ideas, generate questions that deepen understanding, and suggest practical applications.

This assistant becomes a consistent place to process what you learn. Instead of ideas fading after consuming them, they become part of a structured learning system.

Tools such as Readwise, Notion AI, or Mem AI can complement this process. Highlights from books or articles can be collected automatically and later analyzed by AI to identify the themes in what you are learning.

How to Customize an Assistant

Creating such assistants usually involves three steps.

1. **Define the role clearly**: Tell the AI what kind of assistant it should be and what its purpose is.

2. **Provide context**: Share information about goals, routines, and preferences so the assistant understands the environment in which it is working.
3. **Describe how you want it to behave**: For example, you might ask it to:
    - Ask clarifying questions before giving advice.
    - Challenge weak assumptions.
    - Prioritize practical next steps.

These instructions can be saved in many AI platforms so that the assistant behaves consistently each time it is used. Some tools allow for the creation of fully customized assistants or "custom GPTs" that remember these instructions automatically.

Building Systems Around Assistants

Once you begin using assistants regularly, they naturally turn into systems.

- A weekly planning assistant might help review your calendar and identify priorities.
- A reflection assistant might help you review the week every Sunday evening.
- A learning assistant might process notes from books or podcasts.

Each handles a small part of the mental workload.

This idea connects closely with something I explore in my book *Life by Design: Automate to Master Time, Health, Money, and Unlock Personal Success*. That book focuses on designing systems that simplify parts of life that tend to consume time and attention. AI expands those possibilities because many such systems can now be built quickly with simple prompts and tools. When you start thinking this way, you begin to notice opportunities everywhere.

The next time you encounter a recurring task or decision, consider whether it could become a small assistant instead. You might start with something simple:

- Identify a situation you deal with repeatedly.
- Ask AI how it could help analyze or structure that situation.
- Write a short set of instructions that define the assistant's role.

These assistants will begin to add up. A planning helper here, a learning partner there, or a coach that challenges your thinking are dramatic on their own; together, they create a system that supports how you want to live.

Takeaway: Instead of using AI only for occasional tasks, design small custom assistants that help you think, learn, and make decisions more intentionally.

My Personal Experience

Using ChatGPT as a coach has significantly improved my understanding of myself, my business, and my priorities. It has also helped improve my fitness and overall efficiency. Over the past couple of years, I've created three different AI "coaches" that I use extensively:

- A business coach,
- A life coach, and
- A fitness coach.

To create them, I wrote a short set of instructions that described the role I wanted the AI to play, and the type of advice I was looking for, which essentially gave the assistant a "job description."

I use the business coach to help think through ideas for my HomeTechHacker business. I share information about goals, the current state of the business, and ideas I'm considering. From there, the AI helps evaluate which projects might have the greatest impact and which may not be worth pursuing right now. I also use it to brainstorm new ideas, evaluate progress on existing projects, and sometimes help outline the next steps to implement them. The more context I've provided, the more useful the assistant has become.

The life coach has been a different exercise. I asked the AI to act as a brutally honest but trusted advisor and to ask questions that would help it better understand who I am and what I want out of life. I then began asking reflective questions.

AI Prompts:

```
What are my blind spots?

How do you think other people might perceive me?
```

> What changes would you recommend if my goal is to increase my happiness and sense of purpose?

Those conversations helped me step back and think more deliberately about where I am in life and where I want to go—both professionally and personally.

The fitness coach is the assistant I use most often. I originally set it up to support my marathon training. I share my goals and workout results, including distance, pace, heart rate, and how the run felt. The AI then analyzes patterns and offers suggestions for future workouts.

That assistant evolved into a broader health advisor. I now use it to review training trends, think through recovery strategies, analyze lab results, and even generate questions I may want to discuss with my doctors. I interact with this assistant every day, and my fitness and health have improved noticeably since I started using it.

BRINGING IT TOGETHER

Personal growth has always required effort, reflection, and consistency. What has changed is how much support can now be built around that process. Instead of relying entirely on memory, motivation, or scattered notes, AI allows the creation of systems that help guide progress.

In this chapter, I covered how AI can support several parts of that journey. It can help design habits that are small enough to sustain, act as a reflection partner when motivation fades, help process what you learn so ideas turn into understanding instead of fading away, and help organize competing ideas and identify what deserves focus right now. Most importantly, AI can become part of systems built to manage your life more intentionally. When AI helps structure habits, learning, reflection, and decision-making, the technology stops being just a tool occasionally used and becomes support for your growth.

None of this replaces effort. You still have to decide what matters, show up consistently, and take action. What AI can do is help along the way. It can help clarify thinking, surface patterns that might be missed, and keep useful information organized, so progress, rather than logistics, can be the focus. The goal is to design a few systems that support the direction chosen. When used this way, AI becomes less about productivity and more about intentional living. It helps turn scattered effort into steady progress.

At the same time, the more AI is integrated into thinking, decisions, and personal systems, the more important it becomes to understand how these tools work and what responsibilities come with using them. Before building your long-term AI toolkit, it is worth pausing to consider questions of privacy, data, and ethics that shape how this technology should be used. Those issues are the focus of the next chapter.

AI TOOLS REFERENCED

- **ChatGPT (or similar LLM assistants)**: Used as a thinking partner for habit design, reflection, learning summaries, idea organization, and building custom personal assistants.
- **Notion AI**: A note-taking and workspace platform that can summarize notes, extract action items, organize ideas, and analyze journal entries or learning material.
- **Mem AI**: A knowledge management tool that connects notes automatically and uses AI to surface related ideas and patterns across your notes.
- **Readwise / Readwise Reader**: Tools that collect highlights from books, articles, and documents and resurface them later so key ideas remain visible.
- **Day One / Journey**: Digital journaling applications that help maintain a daily reflection habit and organize long-term personal entries.
- **Voice transcription tools (such as built-in phone dictation or AI transcription apps)**: Allow spoken thoughts to be converted into written notes that can be summarized or organized by AI.
- **Zapier / Make**: Automation platforms that connect different apps and services, allowing AI-generated outputs to trigger workflows, such as saving notes, organizing research, or updating task lists.
- **Custom AI Assistants (Custom GPTs or similar tools)**: Configurable AI assistants that can be tailored to specific roles, such as executive coach, fitness coach, learning assistant, or planning partner.

CHAPTER 7

Privacy and Ethics in AI

By now, you've seen how artificial intelligence can help solve everyday problems. It can help you write more clearly, organize ideas, improve health routines, learn things faster, and even build systems that will simplify life. For many people, these tools quickly became everyday items. That usefulness is part of what makes AI so powerful.

But powerful tools always come with trade-offs. The same systems that can summarize a book, help plan a business idea, or analyze a training session run on massive amounts of data, which requires enormous computing resources, and the technology is evolving quickly.

Some practical questions worth understanding before AI becomes deeply integrated into your daily life include:

- What happens to the information you type into an AI tool?
- Where did these systems learn what they know?
- How reliable are the answers they generate?
- And what responsibilities come with using technology that can produce convincing text, images, and even video?

These questions are not reasons to avoid AI; however, the more useful a technology becomes, the more important it is to understand how it works and how to use it thoughtfully. Artificial intelligence is no different.

In this chapter, I'll look at several of the most important issues surrounding AI so you can better understand how these systems work and how to use them responsibly, including:

- How AI systems are trained and how they use data.
- The ongoing debates around copyright and training datasets.
- Why AI systems sometimes produce confident but incorrect answers, and how patterns in training data can sometimes introduce bias.
- The growing environmental impact of large AI models.
- The ethical questions raised by technologies like deepfakes and synthetic media.

I'll conclude with practical guidance for using AI responsibly in your work and daily life.

My goal is to provide enough understanding so that you can make thoughtful decisions about these tools and what information to share with them. AI can be an incredibly powerful assistant, though it works best if you understand both what it can and can't yet do.

HOW AI SYSTEMS USE DATA

One of the most common questions people have about AI tools is simple: "What happens to the information I type into them?"

To answer that, it helps to understand a little about how modern AI systems work.

Large language models are trained on enormous collections of text and other data. The training process happens long before typing your first prompt. Engineers feed the model vast datasets that may include books, articles, websites, public documents, code repositories, and other written material. The model analyzes patterns in that data and learns statistical relationships between words, phrases, and ideas.

It's important to understand that these systems do not store or memorize entire books or articles. Instead, they learn language patterns. During training, the model repeatedly tries to predict the next word in a sentence based on the words preceding it. This process allows the system to generate responses that sound natural and often appear highly knowledgeable.

Once a model is trained, the process you experience as a user is inference. After typing a prompt into an AI tool, the system analyzes the text and generates a response based on the patterns it learned during training.

This is where privacy questions arise.

What Happens to the Information You Enter?

AI tools handle user data in customized ways. Some systems store prompts and responses to help improve their models. Others allow users to disable such data collection. Enterprise versions of AI tools often include stricter privacy protections and agreements that prevent companies from using prompts for training.

Because policies vary across platforms and may change, it is always a good idea to review the privacy settings and terms of service for the specific tool being used.

A Simple Rule of Thumb

A practical way to think about AI privacy is: Don't paste information into an AI tool that you wouldn't be comfortable sharing with a trusted third party.

That includes things like:

- Confidential business information,
- Private financial details,
- Personal identification numbers,
- Medical records, and
- Proprietary documents or intellectual property.

For many everyday uses, this guidance is easy to follow. Asking an AI tool to summarize a public article, brainstorm ideas, or help rewrite a paragraph generally carries little risk. Concerns arise when people begin sharing sensitive or confidential information without realizing that it may be stored or processed by the service provider.

AI Tools Inside Organizations

Many companies are now adopting AI tools internally, which introduces another layer of privacy considerations.

Organizations often use enterprise versions that include stronger privacy guarantees, data handling agreements, and internal policies about what information can be shared with AI tools. In some cases, companies deploy AI models within internal secure environments so that prompts and data never leave their infrastructure.

If you use AI at work, it's important to follow your organization's guidelines. Many companies are still developing these policies as they learn how to integrate AI responsibly.

A Technology Still Evolving

Data privacy practices around AI are evolving rapidly. Governments are developing new regulations, companies are refining policies, and users are becoming more aware of how their data may be used.

The key takeaway is that AI systems can be incredibly helpful, but they are still services operated by organizations that process data. Understanding that relationship helps you decide what information is appropriate to share and what should remain private.

WHERE DID AI LEARN ALL THIS?

If you've ever used an AI tool and been surprised by how knowledgeable it seems, you might have wondered where that knowledge came from. The answer is in how LLMs are trained. As discussed in the previous section, these systems learn from extremely large datasets containing text, code, and other publicly available information. The goal is to expose the model to enough examples of language that it can recognize patterns and generate useful responses.

But that training approach has raised an important question: "What kinds of data should be used to train AI systems, and who owns that information?"

This issue has become one of the most widely debated topics in the AI field.

Lawsuits and Ongoing Debate

In recent years, several authors, media organizations, and artists have argued that their work was included in AI training datasets without permission. Some have filed lawsuits claiming that AI companies used copyrighted material to train their models.

For example, the *New York Times* filed a lawsuit against OpenAI and Microsoft in 2023, alleging that its articles were used to train AI systems and that the resulting models could reproduce portions of that content. The newspaper argued that this could undermine its business model by allowing AI systems to generate information derived from its reporting.

Similarly, groups of authors have brought lawsuits arguing that their books were included in training datasets without their consent. Visual artists have raised comparable concerns about image-generation systems trained on large collections of artwork and photographs.

These cases are still working their way through courts, and the legal landscape surrounding AI training data is evolving quickly. Courts will likely play a major role in determining how copyright law applies to machine learning models.

Why the Issue Is Complicated

At first glance, the debate may seem straightforward: If copyrighted material was used in training datasets, shouldn't permission be required?

The complication is that training an AI model does not involve copying or storing entire books, articles, or artworks inside the system. Instead, the training process extracts patterns from the data. The model learns how language works, how ideas relate to one another, and how certain topics are typically discussed.

Supporters of current training approaches argue that this process is similar to how people learn. A writer may read thousands of books over a lifetime and develop their own style based on what they have absorbed. In that sense, the AI system is learning patterns rather than memorizing content.

Critics argue that the scale of AI training makes the comparison imperfect. While an individual might read hundreds of books, an AI model may be trained on millions of documents. Some creators believe their work is contributing value to AI systems without compensation or attribution.

Because these questions sit at the intersection of technology, copyright law, and economics, they are unlikely to be resolved quickly.

What This Means for Everyday Users

For most people using AI tools, these debates do not affect how the tools function. You can still use AI to summarize information, brainstorm ideas, or improve your writing. However, the discussion highlights an important principle: AI systems are built on large amounts of human-created knowledge.

When you use AI to generate text, images, or other content, it's worth remembering that the system's capabilities ultimately come from patterns learned across the work of countless people.

That awareness can guide us in using these tools responsibly. For example, if AI helps you produce something that draws heavily on someone else's work, it may still be appropriate to cite sources, give credit, or verify information before sharing it publicly.

The legal rules surrounding training data will continue to evolve. New regulations, licensing agreements, and business models may emerge as courts and policymakers respond to these issues.

In the meantime, understanding the debate helps put AI into context. These systems are powerful because they have learned patterns from the vast body of information humans have produced.

WHEN AI GETS THINGS WRONG

One of the most surprising aspects of AI systems is how confident they sound. Ask a question, and the response arrives in clear, well-structured language that reads as if it is coming from a subject-matter expert.

Most of the time, that response may be helpful and reasonably accurate. But sometimes the system produces an answer that seems convincing, while being partially or completely incorrect. In the AI world, these errors are often referred to as hallucinations.

The term doesn't mean the system is imagining things in a human sense. Instead, it reflects the way LLMs generate responses. Because the model is designed to predict likely sequences of words based on patterns in data, it sometimes produces information that *looks* correct even when the underlying facts are inaccurate.

Why Hallucinations Happen

At a technical level, AI models are not retrieving facts from a database the way a search engine does. Instead, they are generating responses based on probabilities learned during training.

When the model has consumed many examples of a topic during its training, its responses are often reliable. But when a prompt asks about something obscure, ambiguous, or poorly represented in its training data, the model may fill in the gaps with language that seems plausible but isn't correct.

For example, an AI tool might:

- Invent a citation to a research paper that does not exist.
- Misattribute a quote to the wrong person.
- Confidently describe a feature in a product that doesn't exist.
- Provide outdated information about a law, policy, or technical standard.

These kinds of mistakes happen because the model's primary goal is to produce a coherent response, not to guarantee factual accuracy.

When Accuracy Matters Most

For casual uses, hallucinations may not cause serious problems. If asking AI to brainstorm dinner ideas or help rewrite a paragraph, small inaccuracies are unlikely to matter.

But when the stakes are higher, verification becomes important.

Examples include:

- Financial or legal guidance,
- Medical information,
- Technical instructions that affect safety, and
- Academic research or published writing.

In these situations, AI should be treated as a starting point for exploration and not as a final authority.

Practical Ways to Verify AI Output

Fortunately, there are simple habits that dramatically reduce the risk of relying on incorrect information.

- **Use AI to generate ideas or explanations**: But confirm important details using reliable sources. A quick search, official documentation, or reputable publication can often validate the information.
- **Ask AI to provide sources or explain its reasoning**: While this does not guarantee accuracy, it can reveal whether the system is relying on general knowledge or generating unsupported claims.

- **Cross-check answers**: Ask a question differently or use an additional tool. If two systems produce similar explanations, the information is more likely to be reliable.
- **Rely on your own judgment**: If something sounds surprising or too certain, that's usually a signal to pause and verify.

How to Think About AI

A useful mental model is to treat AI the way you might a knowledgeable assistant who works quickly but occasionally makes mistakes. The assistant can help research topics, organize ideas, and draft explanations. But you still review the work before acting on it.

THE HIDDEN COST OF INTELLIGENCE

When you type a question into an AI system and receive an answer in a few seconds, the experience feels almost effortless. From the user's perspective, it can seem like the response appears out of nowhere. Behind the scenes, however, an enormous amount of computing power is involved.

Modern AI systems run within massive data centers filled with specialized graphics processing units (GPUs) or similar AI-focused chips. These processors perform the complex mathematical calculations required to train large models and generate responses to user prompts. Companies, including NVIDIA, AMD, and Google, have developed hardware specifically designed to accelerate this type of computing.

Training a large AI model can require thousands of these processors running simultaneously for weeks or even months. Even after a model is trained, every interaction with users requires additional computation as the system analyzes prompts and generates responses. All this computation requires energy.

The Scale of AI Data Centers

Many of the largest AI systems operate in hyperscale data centers operated by major technology companies. These facilities contain tens of thousands of servers that work together, supported by extensive networking equipment, storage systems, and cooling infrastructure.

According to organizations like the International Energy Agency and major technology publications, global data center electricity consumption has been rising steadily, and the rapid growth of AI workloads is one of the factors contributing to this.

In addition to electricity, large data centers also require cooling systems to prevent servers from overheating. Some facilities rely on advanced liquid cooling technologies, while others use significant amounts of water as part of their cooling processes.

This doesn't mean that using AI is environmentally irresponsible, but it does highlight that the technology has a physical footprint. Every query, image generation, or model training run consumes a small amount of energy somewhere in the world.

Efficiency Is Improving

Fortunately, the tech industry is working aggressively to improve efficiency. Hardware manufacturers are designing more energy-efficient processors. Data center operators are investing heavily in renewable energy sources. Many companies are also optimizing AI models to deliver similar performance while requiring less computation.

These improvements matter because AI adoption is growing rapidly. The more efficient the infrastructure becomes, the more sustainable widespread AI use can be.

A Balanced Perspective

It's also helpful to keep the broader context in mind. Many technologies we rely on every day require large amounts of computing power. Streaming video, cloud computing, cryptocurrency mining, and online gaming all contribute to global data center energy consumption. AI is part of that same ecosystem.

At the same time, AI may also help reduce energy use in other areas. Researchers are already using machine learning to improve power grid efficiency, optimize transportation networks, and design more energy-efficient buildings.

Like many technologies, AI's environmental impact will likely involve trade-offs. The systems require significant resources to operate, but they may also enable innovations that reduce energy consumption elsewhere.

Using Technology Thoughtfully

The meaningful takeaway is simply awareness. Technologies that feel purely digital often rely on large physical infrastructures behind the scenes. Understanding this reality encourages a thoughtful approach when using powerful computing systems.

DEEPFAKES AND SYNTHETIC MEDIA

One of the most impressive capabilities of modern AI systems is their ability to generate content that looks and sounds remarkably real. AI can now create written text, realistic images, convincing voices, and even full videos that appear authentic at first glance.

These creations are often referred to as synthetic media, meaning content generated by a computer rather than captured directly from the real world. One specific type of synthetic media, known as deepfakes, uses AI to mimic the face, voice, or actions of a real person in a fabricated video or audio recording.

The technology behind these systems has improved rapidly in recent years. Tools that once required advanced technical skills are now available through consumer applications that can generate images, clone voices, or create video clips with only a few prompts.

In many cases, these capabilities are used for creative and legitimate purposes. Filmmakers use synthetic voices to recreate historical dialogue, game developers generate realistic characters, and content creators use AI-generated visuals in presentations, marketing materials, and educational content. Unfortunately, the same technology is being misused.

The Risks of Synthetic Media

The primary concern surrounding deepfakes and synthetic media is that they can be used to create convincing but false content.

For example, AI tools can generate:

- Videos that appear to show a public figure saying something they never said,
- Audio recordings that mimic someone's voice, and
- Realistic images that depict events that never occurred.

In the wrong hands, these capabilities can be used to spread misinformation, impersonate individuals, or manipulate public opinion.

Several high-profile incidents have already demonstrated these risks. In some cases, criminals have used AI-generated voices to impersonate company executives in phone calls to employees, attempting to authorize fraudulent financial transfers. News organizations have also reported on deepfake videos circulating online that falsely portray politicians or celebrities.

Another growing concern is non-consensual synthetic media, particularly sexually explicit or pornographic content created using the likeness of real people without permission. These deepfakes often target women and can range from manipulated photographs to fully fabricated videos. For victims, the harm can be severe, including reputational damage, harassment, emotional distress, and loss of privacy. As AI tools become easier to use, the ability to create convincing but entirely false intimate content has become a serious ethical and legal concern.

Because synthetic media can appear highly convincing, it can make it harder to distinguish between authentic content and fabricated material.

Why This Matters for Trust

Historically, photos, videos, and audio recordings have been understood as strong evidence of what had happened in the real world. As synthetic media becomes more sophisticated, that assumption becomes less reliable.

This does not mean that all digital media should suddenly be treated as untrustworthy. However, it does mean that a healthy level of skepticism is becoming more important. When encountering dramatic or surprising content online, especially if it spreads quickly on social media, it is worth asking a few questions:

- Where did this content originate?
- Has it been reported or verified by credible sources?
- Are there signs that it may have been edited or generated?

Developing these habits helps maintain a healthy information environment in a world where powerful media generation tools are widely available.

Responsible Use of AI Media Tools

The same technologies that raise concerns about misinformation can also be used responsibly and creatively. Many people now use AI-generated images or voices in legitimate contexts such as design work, educational materials, or entertainment. When using these tools, a few simple practices help maintain transparency and trust.

If you create AI-generated content, consider clearly indicating that it was generated with AI. Transparency helps avoid confusion and prevents audiences from assuming something is real when it is not.

It's also important to avoid generating content that impersonates real individuals in ways that could mislead others or harm their reputation. Even when technically possible, some uses of AI media cross ethical lines that most people would prefer not to encourage. As with many technologies, the tools themselves are not inherently harmful. What matters is how people choose to use them.

USING AI RESPONSIBLY

Artificial intelligence is becoming easier to use and more powerful every year. As you've seen throughout this book, these tools can help solve real problems, simplify daily tasks, and even support personal growth. However, the same capabilities that make AI useful also make it important to use it thoughtfully.

Most ethical concerns about AI do not arise from people intentionally trying to misuse the technology. Instead, they often stem from using powerful tools without fully thinking through their implications. A few simple habits can go a long way toward ensuring that AI remains a helpful and responsible part of your life.

Be Thoughtful About the Data You Share

One of the most important guidelines is to treat AI systems as services operated by third parties. Even if the interface feels like a private conversation, your prompts may still be processed, stored, or reviewed depending on the platform. Therefore, it is wise to avoid sharing sensitive information such as confidential business data, private financial details, personal identification numbers, or medical records. If you wouldn't feel comfortable sending the information to a trusted professional service or storing it in a cloud application, it is best not to paste it into an AI prompt.

Verify Important Information

As discussed earlier, AI systems sometimes produce responses that sound authoritative but contain inaccuracies. Treating AI as a starting point rather than a final authority helps avoid this problem.

When information affects important decisions, verify it using reliable sources. Official documentation, reputable publications, and subject-matter experts remain essential references. AI is excellent for generating explanations, summaries, and ideas, but human judgment remains the final checkpoint.

Give Credit When It's Due

AI can help generate text, images, and ideas quickly; however, those outputs are still derived from patterns learned from human-created information. If using AI to assist with writing, research, or creative work, it is good practice to remain transparent about that assistance when appropriate. In professional or academic settings, this may include acknowledging AI tools as part of the workflow.

Similarly, if using AI to summarize or explain someone else's work, citing the original source is important.

Watch for Bias

AI systems learn patterns from data created by humans. That data reflects the full complexity of human history, culture, and society, including biases that exist in the real world. Consequently, AI systems can sometimes reproduce or amplify those same patterns.

Researchers have documented several examples. In earlier studies of hiring-related AI tools, some systems were found to favor language that appeared more frequently on résumés from certain demographic groups. In other cases, image-generation tools that have been instructed to create images of professionals, such as doctors or executives, sometimes produce results that disproportionately reflect certain genders or ethnicities. These patterns usually weren't the result of intentional design choices. Instead, they reflected biases present in the underlying training data.

This doesn't mean AI systems are inherently unfair or unusable. It does mean they should be treated as tools that require human judgment. When AI is used to generate ideas, summarize information, or assist with writing, bias may not be a major concern. But when AI systems are used in contexts affecting real people—such as hiring decisions, lending evaluations, or policy analysis—human oversight becomes essential.

Avoid Overreliance

AI can be extremely convenient. It can draft emails, summarize documents, and generate ideas in seconds. However, convenience can sometimes lead to overreliance. Maintaining your thinking skills, creativity, and decision-making ability remains important. AI should support thinking, not replace it.

Use AI in Ways That Build Trust

Finally, how AI is used contributes to the broader trust people will place in the technology. Using AI to deceive others, impersonate individuals, or create misleading content undermines trust. Using AI transparently and responsibly helps maintain a healthier digital environment.

As AI tools continue to spread across workplaces, schools, and everyday life, responsible users will play an important role in shaping how they are perceived and adopted.

Fortunately, the ethical use of AI rarely requires complicated rules. Most of the time, it comes down to the same principles that guide responsible behavior in other areas of life: honesty, respect for others, and thoughtful decision-making.

BRINGING IT TOGETHER

Throughout this book, I've explored the many ways artificial intelligence can help solve everyday problems. From improving communication to organizing family life, learning faster, and building systems that support personal growth, AI can be a remarkably useful tool.

But usefulness alone is not the full story. Like any powerful technology, AI works best when both what it can do and the responsibilities that come with using it are understood, including:

- How these systems are trained and how they use data.
- The ongoing debates around copyrighted training material and the lawsuits that may shape the future of AI development.

- The reality that AI can sometimes produce confident but incorrect answers, and why verification remains important.
- The environmental impact of the large data centers that power these systems.
- The ethical concerns raised by synthetic media and deepfakes.

None of these issues means AI should be avoided. Instead, they reinforce the importance of using it thoughtfully. In many ways, AI is like other technologies that have reshaped daily life. The internet, smartphones, and cloud computing all introduced new capabilities along with new responsibilities. People learned how to use those technologies safely and productively.

To use them wisely, it's important to understand how these systems function. When that awareness is combined with the practical techniques covered in this book, AI becomes something far more valuable than just a novelty. It becomes a tool you can intentionally incorporate into your life.

CHAPTER 8

Your Personal AI Toolkit and What's Next

Throughout this book, I've explored many practical ways artificial intelligence can help solve everyday problems. But understanding what AI can do is only the first step. The real advantage comes from learning how to incorporate these tools into your daily routines in ways that make life easier, more productive, and more intentional. Just as the internet or smartphones before it, AI is quickly becoming part of the normal toolkit people use to navigate work, relationships, and everyday decisions.

You don't need to become an AI expert to benefit from it. In fact, the people who gain the most from these tools are often those who focus less on the technology and more on the problems they want to solve. It's empowering to approach AI from the angle of building a small set of capabilities that consistently help you think, create, and act more effectively. That's where the idea of a personal AI toolkit comes in.

A personal AI toolkit is simply the collection of tools, habits, and work-flows that are utilized to make AI a useful part of your life. For some people, that might mean using AI to help with writing, learning, or planning. For others, it may involve automating repetitive tasks, synthesizing information, or building systems that simplify everyday decisions.

Building a toolkit doesn't require mastering dozens of tools or constantly chasing the latest trend. A small number of well-chosen products, combined with the practical techniques from this book, can make a meaningful difference.

Artificial intelligence is evolving quickly; the tools available today will continue to improve. The most important skill is learning how to use these technologies thoughtfully, so they consistently help you focus time, energy, and attention on the things that matter most.

BUILDING YOUR PERSONAL AI TOOLKIT

When people first start exploring AI tools, it's easy to feel like you need to try everything. New tools appear almost daily, each promising to revolutionize productivity, creativity, or automation. However, building a useful AI toolkit involves selecting a small set of tools that will consistently help solve real-life problems. Keeping up with every new AI application is not necessary.

Think of it like kitchen or mechanical tools. You probably don't own every possible tool, but the ones you do are used regularly. A few reliable tools, used well, can accomplish far more than a large collection that is barely touched. An AI toolkit should work the same way.

Most people find that their toolkit naturally falls into a few simple categories, each of which helps solve a different type of problem; together, they create a flexible system you can use across many areas of life.

Thinking Partners

The most common entry point into AI is using it as a thinking partner. Tools like ChatGPT, Claude, and Gemini are especially useful for this because they allow you to interact conversationally while exploring ideas, solving problems, or organizing your thoughts.

A thinking partner AI can help you:

- Brainstorm ideas when you feel stuck,
- Organize complex information,
- Explain difficult concepts in simpler terms,
- Evaluate decisions or trade-offs, and
- Outline projects or plans.

Throughout the book, I've presented many such examples. When you ask AI to summarize information, draft messages, prioritize ideas, or break down a complicated topic, you're essentially using it as an external thinking partner.

Many people find that this is what they use most frequently. Instead of staring at a blank page or trying to untangle a complicated problem alone, you now have a tool that can help you think ideas through faster.

Productivity Assistants

Another powerful category of tools helps with everyday productivity tasks, integrating AI directly into the software people already use for work, writing, or organizing information.

Examples include:

- Microsoft Copilot in Word, Excel, and Outlook;
- Google Workspace AI tools in Docs and Gmail; and
- Notion AI for organizing notes and projects.

These tools are especially good at tasks like:

- Summarizing long documents,
- Drafting emails or reports, and
- Turning rough ideas into structured outlines.

For example, instead of manually reviewing a long meeting transcript or document, you can ask an AI assistant to summarize the key points and highlight decisions that were made. Instead of starting from scratch when writing a report, you can provide bullet points and ask the tool to expand them into a draft. Used thoughtfully, these tools remove friction from routine tasks, making it more likely you can focus on ideas and decisions that matter more.

Information Capture

One of the most overlooked parts of building a personal AI toolkit is how to capture information in the first place. AI can only help you think, organize, and act on information that it has access to. If ideas, meetings, and conversations dissipate, the system breaks down before it even starts. This is where AI-powered note-taking tools come in; they can capture information automatically and transform it into something that can be used later.

For meetings, tools like Zoom AI Companion, Microsoft Teams Copilot, Google Meet AI, and Otter.ai can record conversations, generate summaries, and identify action items. After a meeting, instead of starting from scratch, you already have a structured summary with key points and follow-ups. This is especially useful when conversations move quickly or involve multiple people.

For personal notes, tools like Notion AI, Evernote with AI search, and Obsidian with AI plugins allow you to capture ideas and retrieve them later using natural language. Instead of trying to remember where you wrote something down, you can interface with them using phrases like "Summarize my notes from last week about content ideas" or "What decisions did I make about my website redesign?"

The key is to capture notes in a way that AI can work with, meaning:

- Capturing consistently (don't rely on memory),
- Keeping information in one or two primary systems, and
- Using tools that allow search, summarization, or interaction.

You don't need a complex setup. A simple system works well: use an AI meeting tool to capture conversations, a central note tool to store ideas, and an AI assistant to summarize and connect everything.

When a capture system is working, AI tools become significantly more valuable. You're no longer starting from a blank page, but building upon a record of your thinking, conversations, and decisions.

Creative Tools

AI is also rapidly expanding what individuals can do creatively. Tools like DALL-E, Midjourney, Runway, and Adobe Firefly allow people to generate images, graphics, videos, and design elements with simple text prompts.

Even if you don't consider yourself a designer or creative professional, these tools can still be useful in everyday situations. For example, you might use AI creative tools to:

- Generate images for presentations or social media.
- Create visual concepts for a project or website.
- Produce simple marketing graphics.
- Experiment with design ideas before committing to them.

In the past, such tasks required specialized skills or expensive software. AI tools are lowering the barriers, allowing individuals and small teams to create high-quality visual content much more quickly.

Automation and Workflows

The final category of tools focuses on automation. Instead of simply helping you think or create, these tools help connect different systems and automate repetitive tasks.

Platforms like Zapier and Make allow you to build automated workflows between apps. When combined with AI tools, they can create powerful systems that quietly handle routine tasks in the background.

For example, you might build workflows that:

- Summarize incoming emails and send you daily digests,
- Organize documents automatically into folders,
- Create summaries of meeting transcripts, and
- Send reminders or updates when certain events happen.

These kinds of automations can save small amounts of time repeatedly over the course of a week, which will significantly reduce mental load and repetitive work.

I explore this idea—designing systems that simplify everyday tasks—more deeply in my book *Life by Design: Automate to Master Time, Health, Money, and Unlock Personal Success*. AI tools are increasingly becoming a powerful part of such systems, making it easier than ever to automate small pieces of your life in ways that previously required technical expertise.

As you begin building an AI toolkit, don't adopt tools in every category immediately. A simple starting point might look like this:

- One thinking partner AI tool,
- One productivity assistant integrated into your daily work,
- One creative tool for generating visuals or ideas, and
- One automation platform to simplify repetitive tasks.

Before long, you'll discover which tools fit best into routines and which ones are rarely used. Your toolkit will evolve as the technology evolves. Aim to build a small set of tools that reliably help you think more clearly, work more efficiently, and spend more of your time on the things that matter most.

THE FUTURE OF AI IN DAILY LIFE

If there's one lesson from the history of technology, it's that tools rarely stay separate from our daily lives for long. The internet once required dialing into special services. Smartphones were once optional gadgets. Today, they are simply part of how the world works. Artificial intelligence is following a similar path.

Currently, many people think of AI as something you open in a browser or an app. You go to a chatbot, type a question, and get an answer. Eventually, AI will become less of a destination and more of an invisible capability built into the tools and devices you already use every day.

Many software platforms are already integrating AI directly into their products. Instead of visiting a separate AI tool, its capabilities appear within applications.

For example, productivity software, like Microsoft Office and Google Workspace, now includes AI assistants that can summarize documents, draft emails, and organize information. Search engines are evolving from simply listing links to generating answers and explanations. Even operating systems and web browsers are beginning to incorporate AI features that help organize files, manage notifications, and automate tasks.

This trend will continue. AI will quietly assist with everyday tasks, such as writing messages, planning schedules, analyzing information, and finding answers faster. Many people may not even think of themselves as "using AI." They will simply experience software that is more helpful and responsive.

Personal AI Assistants Will Become More Capable

Another major change will be the improvement of personal AI assistants. Today's AI tools already act as useful thinking partners, but future versions will likely become more integrated into daily routines.

More assistants will connect to calendar, documents, notes, and other systems you already use. Instead of just answering questions, they will help plan projects, track goals, and manage information more effectively.

Imagine an assistant that can review your schedule for the week and suggest better ways to organize your time. Or one that can help summarize research gathered for a project and highlight the most important insights. As mentioned earlier, such capabilities are already beginning to appear in early forms, and they will likely become more refined.

The key point is that these tools will become increasingly personalized. They will learn how you prefer to work, what kinds of information you value, and how to present suggestions in ways that are most useful to you.

AI Will Help Automate More Routine Tasks

Another important shift will involve automation. Today's AI systems mostly respond to prompts or questions, but the next generation of tools will increasingly be able to carry out sequences of actions. Sometimes referred to as AI agents, many are already here. Instead of simply providing information, they complete tasks while you supervise the results.

For example, an AI assistant might gather research on a topic, summarize the key findings, and draft an outline for a report. It might organize documents, monitor information, or help coordinate small projects. These capabilities are still in development, but they point toward a future where AI tools assist not just with thinking but also with execution.

Even as these systems become more capable, human judgment will remain essential. AI can help accelerate work and reduce routine tasks, but people still provide the context, priorities, and values that guide decisions.

AI Literacy Will Become a Normal Skill

As these tools become more common, knowing how to use them will become a basic life skill. Just as people learned how to search the web, manage email, and navigate smartphones, the next generation of digital literacy will include knowing how to collaborate with AI tools.

This doesn't mean everyone needs to understand the technical side of AI systems. Most people don't understand how the internet works at a technical level either. What matters is learning how to ask the right questions, evaluate the answers, and apply the tools thoughtfully.

In many ways, that's what this book has been about. The techniques I've covered are less about mastering specific tools and more about developing a mindset for using AI as a partner in solving everyday problems.

The technology will continue to evolve. New tools will appear, existing ones will improve, and some platforms will disappear altogether. But the ability to work effectively with AI—to guide it toward useful outcomes and apply it to real problems—will remain valuable, no matter how much the tools change.

WHERE TO GO NEXT

Like any tool, the real value of AI comes from using it consistently. You don't need to master every AI tool. What matters most is developing the habit of turning to AI when faced with a problem that could be solved faster, more clearly, and with less effort.

The simplest way is to start small. Pick one or two areas of life where you often feel stuck, overwhelmed, or slowed down by routine tasks. The next time one of these situations comes up, try involving AI as a thinking partner. For example:

- Ask AI to help outline an idea before you start writing.
- Summarize a long article or report you don't have time to read in full.
- Generate options for solving a problem you're facing.
- Help you plan a project or break a large task into manageable steps.

You'll naturally discover which kinds of prompts and workflows work best. By continuing to experiment, you'll also find new tools and capabilities that expand what AI can help you accomplish.

If you'd like to continue learning about practical ways to use technology to simplify your life, explore additional resources and articles on my website, HomeTechHacker.com. I regularly share guides, tools, and ideas focused on making technology more useful and approachable.

One of the most important things to remember is that you don't need to wait until AI tools are perfect before using them. Every major technology has gone through cycles of improvement and refinement. The people who benefit the most are usually those who start experimenting early and gradually build confidence.

Artificial intelligence is still evolving, but its potential to improve how we think, work, and organize is already clear. Used thoughtfully, these tools can help reduce everyday friction, unlock new ways of solving problems, and open up more time and energy to focus on what matters most.

A PERSONAL AI PLAYBOOK

The most effective way to adopt AI is to start small, experiment often, and gradually build habits around the tools that work best for you.

If wondering where to start, this simple playbook can help guide your first steps:

1. **Choose your thinking partner**: Pick one AI tool and make it your primary assistant. Tools like ChatGPT, Claude, or Gemini are good starting points. Become comfortable working with one.
2. **Start with everyday problems**: AI is most useful when applied to real situations you face regularly. Use it to draft messages, summarize information, brainstorm ideas, or help plan projects. The more practical the problem, the more useful the tool becomes.
3. **Ask better questions**: The quality of AI output often depends on the quality of the prompt. Provide context. Explain what

you want. If the first answer isn't quite right, ask follow-up questions. Treat the interaction like a conversation rather than a single command.

4. **Build small systems**: Look for tasks that are repeated frequently and explore ways AI can simplify them. This might include organizing information, summarizing research, planning schedules, or automating small workflows.
5. **Keep experimenting**: AI tools will continue to improve rapidly. New capabilities will appear, and existing tools will become more powerful. The best way to stay current is to keep experimenting and remain curious about how these tools can help you solve problems.

These small habits will turn into something much more powerful: A personal AI toolkit that helps you think more clearly, move faster with ideas, and reduce the mental load of everyday life.

AI isn't here to replace human thinking. Used well, it does the opposite. It frees up time and attention, so you can focus on the ideas, relationships, and goals that matter most. And that's ultimately the promise of AI when applied thoughtfully: not more technology in your life, but a better life with the help of technology.

Technology has always shaped how we live, but tools like artificial intelligence give us something new: the ability to think alongside machines that can help us organize information, explore ideas, and solve problems faster than ever before. The real opportunity is using these tools to create more space in our lives for the things that matter most.

If AI can help you think more clearly, make better decisions, and spend more time on the people, work, and goals that are meaningful to you, then it has already done something powerful. The tools will continue to change, but the opportunity remains the same: using technology intentionally to build a life that works better for you.

The future of AI isn't about the technology. It's about what you choose to do with it.

GLOSSARY

This glossary explains key artificial intelligence terms used throughout the book in simple language. The goal isn't to make you an AI engineer, but to provide enough understanding so that you can confidently use these tools in everyday life.

AI Agent: A software system that can perform tasks or take actions on your behalf using artificial intelligence. AI agents may gather information, complete steps in a workflow, or interact with other software tools.

AI Assistant: An AI tool designed to help users complete tasks, answer questions, organize information, or support decision-making through conversation or automation.

AI Bias: When an AI system produces results that unfairly favor or disadvantage certain groups of people, it is often because the data used to train the system reflects existing societal biases.

AI Literacy: The ability to understand what artificial intelligence can and cannot do, ask effective questions, interpret results, and use AI tools responsibly.

Artificial Intelligence (AI): Computer systems designed to perform tasks that typically require human intelligence, such as recognizing patterns, understanding language, generating content, or making predictions.

Automation: The use of technology to perform tasks automatically with little or no human intervention. AI can enhance automation by allowing systems to make decisions or adapt to new information. *See also: Workflow Automation.*

Chatbot: A conversational interface that allows users to interact with an AI system by typing or speaking questions and receiving responses in natural language.

Decision Fatigue: The mental exhaustion that comes from making many decisions throughout the day. AI tools can help reduce decision fatigue by organizing information or suggesting options.

Deepfake: Synthetic audio, video, or images created using AI that make it appear that a person said or did something they did not actually say or do.

Ethical AI: The practice of designing and using AI systems in ways that promote fairness, transparency, accountability, and responsible decision-making.

Generative AI: A type of artificial intelligence capable of creating new content such as text, images, music, video, or computer code based on patterns learned during training. *See also: Large Language Model.*

Hallucination (AI Hallucination): When an AI system produces information that sounds convincing but is incorrect or fabricated.

Large Language Model (LLM): A type of AI model trained on massive amounts of text data that can understand and generate human-like language. *See also: Generative AI.*

Machine Learning: A method of building AI systems that learn patterns from data and improve performance without being explicitly programmed for every task.

Natural Language Processing (NLP): A branch of artificial intelligence focused on enabling computers to understand, interpret, and generate human language.

Personal AI Assistant: An AI tool customized to support an individual's specific needs, such as planning projects, organizing ideas, tracking goals, or providing coaching.

Prompt: The instruction, question, or request given to an AI system that guides the response it generates. *See also: Prompt Engineering.*

Prompt Engineering: The practice of writing prompts in ways that produce more useful, accurate, or detailed responses from AI systems. *See also: Prompt.*

Synthetic Media: Content such as images, video, or audio generated or significantly altered by artificial intelligence.

Thinking Partner: A way of using AI as a collaborative tool to brainstorm ideas, organize thoughts, explore solutions, or clarify complex topics.

Training Data: The large collections of text, images, audio, or other information used to teach an AI system how to recognize patterns and generate responses.

Workflow: A series of steps used to complete a task or process. AI tools can help simplify or automate many parts of a workflow.

Workflow Automation: The use of software tools, often combined with AI, to automatically perform repetitive tasks or processes. *See also: Automation.*

THANK YOU

Thank you for purchasing *Your Life, Solved*. I know that you could have picked any number of books to read, but you've picked this book, and for that, I am extremely appreciative.

I hope that it has inspired you and helped you spend time on the things in your life that are most important to you. If so, please share this book with your friends and family by **posting about it on Twitter, Instagram, Facebook, and Pinterest**.

If you enjoyed this book and found some benefit from reading it, I'd love to hear from you. I hope that you can take the time to **post a review on Amazon**. Your feedback and support will help me greatly improve my writing craft for future projects.

I want you, the reader, to know that your review is very important and very much appreciated.

I wish you all the best in your journey to use AI to improve your life!

ABOUT THE AUTHOR

Marlon Buchanan, bestselling author and IT Director, holds a bachelor's degree in Computer Science and Engineering from MIT and master's degrees in Business Administration and Software Engineering from Seattle University. With over 25 years of experience in IT, software development, and smart home automation, he helps busy professionals harness technology to create more time, reduce stress, and design lives they love.

He has been automating things around his house since he was a kid. Now, his wife and kids get to enjoy the fruits of his automation and efficiency exploits. He is best known for his smart home, home networking, and cybersecurity articles on his blog HomeTechHacker.com, and his bestselling books *Life by Design*, *The Personal Cybersecurity Manual*, *The Home Network Manual*, and *The Smart Home Manual*.

Be sure to check out HomeTechHacker.com for free in-depth articles about AI, automation, home networks, cybersecurity, and more. Also, subscribe to his newsletter to receive the latest updates about automating your life and other home technology topics. You can also follow him on these social media channels (@HomeTechHacker):

- X (Twitter): twitter.com/HomeTechHacker
- Instagram: instagram.com/HomeTechHacker/
- Pinterest: pinterest.com/HomeTechHacker
- Facebook: facebook.com/HomeTechHacker

CONSIDER MARLON BUCHANAN'S OTHER BOOKS

- *Life by Design: Automate to Master Time, Health, Money, and Unlock Personal Success*
- *The Personal Cybersecurity Manual: How Anyone Can Protect Themselves from Fraud, Identity Theft, and Other Cybercrimes*
- *The Home Network Manual: The Complete Guide to Setting Up, Upgrading, and Securing Your Home Network*
- *The Smart Home Manual: How to Automate Your Home to Keep Your Family Entertained, Comfortable, and Safe*
- *Home Wi-Fi Tuneup: Practical Steps You Can Take to Speed Up, Stabilize, and Secure Your Home Wi-Fi*

All books are available in e-book and paperback formats. *Life by Design*, *The Personal Cybersecurity Manual*, *The Home Network Manual*, and *The Smart Home Manual* are also available as audiobooks. You can learn more about these books, including where to buy them, at MarlonBuchanan.com.

www.ingramcontent.com/pod-product-compliance
Lightning Source LLC
Chambersburg PA
CBHW071441230926
51846CB00001B/303
9781958648070